Discrete Discriminant Analysis

Applied Probability and Statistics

continued on back

Discrete
Discriminant Analysis

MATTHEW GOLDSTEIN

Baruch College, City University of New York

WILLIAM R. DILLON

University of Massachusetts

John Wiley & Sons

New York • Chichester • Brisbane • Toronto

Library of Congress Cataloging in Publication Data

Goldstein, Matthew, 1941-
   Discrete discriminant analysis.

   (Wiley series in probability and mathematical
statistics)
   Bibliography: p.
   Includes index.
   1.   Discriminant analysis. I. Dillon, William R.,
joint author. II. Title.

QA278.65.G64   519.5'3   78-2899
ISBN 0-471-04167-X

Printed in the United States of America

10 9 8 7 6 5 4 3 2 1

To
Ellyn and *Mary*

Preface

Researchers in the social, behavioral, and biological sciences often collect discrete multivariate data. However, it has been only in the past few years, as a result of the explosion in the computer field, that statisticians have been encouraged to pursue research activity leading to methodologies specifically designed for such data. Without question, the area of greatest concentrated work in the analysis of discrete multivariate observations has been with data that are cross-classified. A comprehensive survey that details most of the important work dealing with the analysis of cross-classified data has recently been published by Bishop, Fienberg and Holland (1975).

Because so many varied fields involve studies emitting discrete observations, the literature dealing with such issues is not concentrated in any one field. In particular, problems in so-called discrete discriminant analysis can be found not only in journals of statistics but also in the literature of biology, sociology, psychology, consumer behavior, and engineering. We decided to write this book because the problem of discrimination between groups with discrete multivariate observations is viewed to be an important one, and because there does not exist a unified source for the applied statistician or quantitative research worker to gain information about the subject.

Although the book is short, it is a comprehensive up-to-date treatment presenting both the theory and numerous applications of discrete discriminant analysis. It is expected that the reader has a sound basic background in ideas in probability and statistics. However, because there are numerous examples that are worked out in detail, readers with less than this background should also be able to gain some feel for the subject matter.

Chapter 1 deals with problems that can arise with the utilization of the standard linear discriminant function applied to discrete data. The discussion for the most part is that of motivation for the consideration of more general classification procedures specifically designed to deal with discrete observations.

Chapter 2 is the main chapter of the text. This chapter discusses models in discrete discriminant analysis and is a comprehensive survey along with theory and many illustrative examples. The chapter is written in such a way that professional statisticians as well as research workers can get both the theoretical justification of what is discussed along with workable methodology for their own particular problems. The examples come from a wide variety of different disciplines chosen specifically to address different interests.

Chapter 3 is principally theoretical and deals specifically with the problems of error rates and bias. The chapter, however, is not overly detailed, and the reader is requested to consult primary sources for additional information if needed.

Chapter 4 is devoted to the problems of variable selection. As we indicate in the chapter, this is a particularly important problem in discrete discriminant analysis because of difficulties encountered with large, sparse data sets. All of the available procedures that presently exist in this area are surveyed in sufficient detail that anyone reading the material should have adequate information to use them. The procedures discussed are illustrated with one particular data set so that comparisons can be made.

Chapter 5 deals with special topics. Here we discuss how classification procedures can be formed when the data consist of both continuous and discrete components. In addition to considering the most up-to-date methods, we also propose some general extensions that have at this particular date not appeared in the literature. Another section in this chapter is devoted to the problem of how one would compare two competing procedures for a given data set. Lastly, a section is devoted to simulation experiments that have appeared in the literature and address the problem of comparative performance of various classification schemes that have been proposed principally on binary data.

Chapter 6 is different in scope in that it is devoted to a discussion of the computer programs that presently exist to implement most of the techniques discussed in the previous chapters of the text. Many of these programs have been written by the authors, and complete documentation is provided so that the user of these methods should have no problem with their implementation.

We wish to thank Mrs. Bonnie Webster for her diligent effort in typing the entire manuscript.

MATTHEW GOLDSTEIN
WILLIAM R. DILLON

New York, New York
Amherst, Massachusetts
March 1978

Contents

Discrete Discriminant Analysis

CHAPTER 1

The Linear Discriminant Function

1.1 INTRODUCTION

Although our efforts focus in principle upon discrete discriminant analysis and problems related to the subject, it might appear a severe omission on our part if we did not devote some discussion to the linear discriminant function (LDF). From a historical perspective, the linear discriminant function can be accepted as the pablum of classification research. Developed in 1935 by Fisher to answer perhaps one of the most fundamental of all systematic problems of the taxonomic variety, it stands as both the first clear statement of the problem of discrimination and the first proposed solution.

Since the introduction of this function, its behavior has been considered from the point of view of its distribution when samples are used to estimate the coefficients by Anderson [1951, 1973] and Okamoto [1963], robustness to various departures from the assumptions for optimality by Cochran and Hopkins [1961], Gilbert [1968, 1969], Lachenbruch [1966], Moore [1973], and Dillon and Goldstein [1978], selection of variates by Cochran and Bliss [1948] and Cochran [1964], and estimation of error rates by Dunn [1971], Hills [1967], and Lachenbruch and Mickey [1968]. Though far from exhaustive, this list is sufficient to attest to the widespread interest and varying degrees of orientation taken in the study of Fisher's LDF. Furthermore, as a survey of the applied research in the social, behavioral, business, and medical sciences would indicate, use of this discriminant function in analysis has been extensive, by the most conservative standards.

Obviously, the communality in applied research is the goal to assign individuals or objects to a specific population group on the basis of data that are related to the group members. However, a natural question that arises when discussing applied research concerns the notion of performance or, in more general terms, the compatibility of data with underlying

1

assumptions for optimality. As is the case for many multivariate techniques, herein lies the problem since, as many applied researchers have come to realize, the situation covered by this discriminant function is in reality rare. Although Fisher's original approach is distribution-free, it provides an optimal assignment rule, in the sense of minimizing the probability of misclassification, only if multivariate normality and equal covariance matrices are assumed present.

To illustrate this, in the next section we take a general approach to the two-group problem and define optimality within this context. The arguments to some extent are repeated in the beginning of Chapter 2, where considerations of optimality are discussed for the discrete problem. Our discussion of the classification rule first derived by Fisher is brief, since this subject matter is adequately presented elsewhere (Anderson [1952]; Lachenbruch [1975]), and since we feel that the primary focus of the chapter should be on those practical, methodological problems arising with the use of Fisher's LDF on the common types of data collected in the social, behavioral, and business sciences. In particular, Section 1.3 comments on the likely anomalies arising with its use in situations where the data are qualitative, each measurement taking on a finite and usually small number of values. We hope, in our attempt to provide some insights into practical considerations, that we may alert the reader as to why researchers should take a closer look at the classification problem rather than simply to assume multivariate normality and then proceed to use a variant of Fisher's LDF.

1.2 FISHER'S RULE

Suppose that two groups Π_1 and Π_2 mix in a large population having prior probabilities δ_1 and δ_2. Hence if Π indexes an individual's group, then $P\{\Pi=\Pi_i\}=\delta_i$, $i=1$ or 2. Let $F_i(\mathbf{x})=P\{\mathbf{X}\leqslant\mathbf{x}|\Pi=\Pi_i\}$ denote the class-conditional distributions and $f_i(\mathbf{x})$ their associated conditional densities, where $\mathbf{X}=(X_1,X_2,\ldots,X_p)$ is a random p-dimensional vector whose realizations generate a sample space $\mathfrak{X}$. The unconditional density at $\mathbf{x}$ is simply expressed as

$$g(\mathbf{x})=\delta_1 f_1(\mathbf{x})+\delta_2 f_2(\mathbf{x})=g_1(\mathbf{x})+g_2(\mathbf{x}),$$

where the $g_i(\mathbf{x})$ are called the *discriminant scores*.

A classification rule D may be thought of as an ordered partition $D=\langle D_1,D_2\rangle$ of $\mathfrak{X}$, where D assigns an individual to group Π_i if and only if $\mathbf{X}=\mathbf{x}\in D_i$. The conditional probability of misclassification given $\mathbf{X}=\mathbf{x}\in D_i$ is $t(D|\mathbf{X}=\mathbf{x})=g_j(\mathbf{x})/g(\mathbf{x})$ for $i\neq j$. For continuous measurements the unconditional probability of misclassification $t(D)$ can be expressed as

$$t(D) = \int_{D_1} g_2(\mathbf{x})\,d\mathbf{x} + \int_{D_2} g_1(\mathbf{x})\,d\mathbf{x} \qquad (1.2\text{-}1)$$

If the collection of all classification rules is denoted by $\mathcal{D}$, then we may define an optimal rule D' as one satisfying

$$t(D') = \inf_{D \in \mathcal{D}} t(D) \qquad (1.2\text{-}2)$$

It is well known (Welch [1939]; Hoel and Peterson [1949]) that a particular optimal partition of $\mathfrak{X}$ is defined by $D^* = \langle D_1^*, D_2^* \rangle$, where

$$D_1^* = \{\mathbf{x}\,|\,g_1(\mathbf{x}) > g_2(\mathbf{x})\}$$

$$D_2^* = \{\mathbf{x}\,|\,g_1(\mathbf{x}) < g_2(\mathbf{x})\} \qquad (1.2\text{-}3)$$

and points $\mathbf{x}$ for which $g_1(\mathbf{x}) = g_2(\mathbf{x})$ are randomly assigned to D, $i = 1,2$.

If we apply this rule to the case where the two underlying densities are multivariate normal with mean vectors $\boldsymbol{\mu}_1$ and $\boldsymbol{\mu}_2$ and common covariance matrix $\boldsymbol{\Sigma}$, then

$$D_1^* = \left\{ \mathbf{x}\,|\,(\boldsymbol{\mu}_1 - \boldsymbol{\mu}_2)'\boldsymbol{\Sigma}^{-1}\left[\mathbf{x} - \tfrac{1}{2}(\boldsymbol{\mu}_1 + \boldsymbol{\mu}_2)\right] \geqslant \frac{\delta_2}{\delta_1} \right\}$$

$$D_2^* = \left\{ \mathbf{x}\,|\,(\boldsymbol{\mu}_1 - \boldsymbol{\mu}_2)'\boldsymbol{\Sigma}^{-1}\left[\mathbf{x} - \tfrac{1}{2}(\boldsymbol{\mu}_1 + \boldsymbol{\mu}_2)\right] < \frac{\delta_2}{\delta_1} \right\} \qquad (1.2\text{-}4)$$

This optimal partition is usually referred to as a *linear discriminant*. A rule equivalent to (1.4) was derived by Fisher [1936], who considered finding the linear function $\boldsymbol{\beta}'\mathbf{x}$ that maximizes the ratio

$$\frac{(\boldsymbol{\beta}'\boldsymbol{\mu}_1 - \boldsymbol{\beta}'\boldsymbol{\mu}_2)^2}{\boldsymbol{\beta}'\boldsymbol{\Sigma}\boldsymbol{\beta}}. \qquad (1.2\text{-}5)$$

Note that this expression is the ratio of the squared mean difference between the two groups to the assumed common variance. In either case, the partition given in (1.2-4) or its equivalent with appropriate maximum likelihood estimates replacing the parameters $\boldsymbol{\mu}_1$, $\boldsymbol{\mu}_2$, and $\boldsymbol{\Sigma}$ is probably the most frequently utilized rule for classification.

If the assumptions of normality and common covariance matrix are present, then the sample-based version (i.e., replacing $\boldsymbol{\mu}_1$, $\boldsymbol{\mu}_2$, and $\boldsymbol{\Sigma}$ by their respective maximum likelihood estimates $\hat{\mu}_1$, $\hat{\mu}_2$, and $\hat{\Sigma}$) of (1.2-4) is still a consistent procedure. However, if the data violate the assumed model assumptions, then the partition given in (1.2-4) will not be optimal. For example, it is well known that if the covariance matrices in Π_1 and Π_2 are quite different, whereas multivariate normality remains intact, then the quadratic discriminant function of Smith [1947] is appropriate. Similarly, if the data violate the normality assumption, then the use of (1.2-4) cannot be justified, at least not from an optimality argument. In the remaining sections we discuss some of the likely anomalies arising with the use of a Fisher-type linear discriminant when the available data are discrete.

1.3   THE USE OF QUALITATIVE DATA

In many fields, especially in the behavioral, social, and business sciences, measurement is still a very real problem. Frequently, the researcher has no meaningful alternative but to settle for categorical dichotomies or imprecise qualities. In much of the applied research, therefore, measurements are commonly collected by the use of two-, three-, or four-point scales, which are at best of a nominal or ordinal nature and for which the obtained data can be characterized as being qualitative. Naturally, a question that immediately appears is how best to analyze such data. It would seem that many applied researchers in cases such as this feel comfortable in ignoring the discrete nature of the data and proceed with continuous variable techniques. Of course, in the context of the classification problem this leads to the use of Fisher's LDF. In particular, common practice before a classification analysis is performed has the researcher assign numeric scores to the levels of the variables or dichotomize the levels and then apply some variant of Fisher's LDF.

As one might suspect, in the vast majority of applied research the application of Fisher's function has not been preceded by tests to determine if the conditions for its optimality are satisfied. The authors are of the opinion that researchers have apparently applied the technique in the hope of obtaining useful if not optimal results. Although one should, of course, be concerned with the underlying statistical questions, we see that our discussion of measurement considerations is applicable in either case.

1.3-1   A Special Case: Binary Predictor Variables

To further motivate the rather unique set of problems arising with the use of Fisher's LDF on qualitative data, we consider first the extreme but common case of dichotomous predictor variables. For illustrative purpose consider the following case study reported by Yerushalmy et al. [1965]. In this study of infant immaturity, the author suggested using birth weight and length of gestation as indices of immaturity, where

$X_1 = 0$ if birth weight is low
    $= 1$ if birth weight is high
$X_2 = 0$ if gestation length is short
    $= 1$ if gestation length is long.

Accordingly, if the length of gestation is short and the birth weight is low or if the length of gestation is long and the birth weight is high, then the newborn child is classified as "normal." On the other hand, "abnormal" babies are those of long gestation but with low birth weight, or those of short gestation but with high birth weight. Thus babies with $x = (0, 0)$ and $(1, 1)$ should be classified as from Π_1, the normal population group, whereas babies with $x = (1, 0)$ and $(0, 1)$ should be classified as belonging to Π_2, the abnormal population group.

Now if a linear procedure is used and if we denote the appropriate cutoff point between Π_1 and Π_2 as Z^*, then $x = (0,0)$ in Π_1 and $x = (0,1)$ in Π_2 implies $\beta_0 < \beta_1 + \beta_2$, or that $\beta_2 > 0$. Similarly, β_1 must be positive if $x = (0,0)$ is classified as belonging to Π_1, and $x = (1,0)$ is classified as Π_2. Therefore, if both β_1 and β_2 are positive, it naturally follows that $\beta_0 + \Sigma_{i=1}^2 \beta_i > Z^*$, and hence $x = (1,1)$ cannot be possibly classified as belonging to the same population group as $x = (0,0)$. In this case we see that any linear procedure is incapable of indicating the (suggestive) proper classification rule regardless of the outcome of any sampling procedure.

Stated more formally, under a linear model the likelihood ratio for any observation x is given by

$$L(\mathbf{x}) = \frac{f_1(\mathbf{x})}{f_2(\mathbf{x})} = \beta_0 + \sum_{j=1}^{p} c_j x_j. \qquad (1.3\text{-}1)$$

It follows, therefore, that for any pair, say x_1 and x_2,

$$L(1,1,x_3,\ldots,x_p) = L(0,1,x_3,\ldots,x_p) + L(1,0,x_3,\ldots,x_p)$$
$$- L(0,0,x_3,\ldots,x_p). \qquad (1.3\text{-}2)$$

If it occurs that

$$L(0,0,x_3,\ldots,x_p) < \min\{L(0,1,x_3,\ldots,x_p), L(1,0,x_3,\ldots,x_p)\} \qquad (1.3\text{-}3)$$

then

$$L(1,1,x_3,\ldots,x_p) > \max\{L(0,1,x_3,\ldots,x_p), L(1,0,x_3,\ldots,x_p)\} \qquad (1.3\text{-}4)$$

and hence no linear model can satisfactorily characterize populations in which the likelihood ratio is nonmonotone. The above arguments were apparently first offered by Moore [1973]. In particular, through Monte Carlo sampling experiments he clearly demonstrated that for certain parameter structures defining the population groups, the log likelihood ratio formed is not monotone with the number of positive X_j. In Moore's language, another way to describe this nonmonotonicity is to say that the likelihood ratio undergoes a reversal.

1.3-2 The General Case of Qualitative Data

As was indicated, one finds two rather heuristic coding schemes increasingly employed in the analysis of qualitative data. In practice, these schemes have the researcher assign numerical values to the levels of the variables or dichotomize the levels such that $k-1$ zero–one dummy variables are created for each variable having k levels ($k-1$ dummy variables are created instead of k to avoid the problem of singularity). In the latter case our discussion of the last section clearly indicated that the use of Fisher's LDF on binary variables can severely affect the accuracy of classification, at least in population structures for which the likelihood ratio undergoes a reversal. In the former case, wherein the variables can assume any finite but generally small number of values, the situation is somewhat more complex. However, from a computational point of view no additional difficulties are created by including in the linear discriminant function qualitative data—nominal and ordinal scaled variables—but the use of such data coding schemes is nevertheless questionable and suffers from a number of deficiencies.

To motivate the discussion, consider for the moment the variable descriptions given in Table 1.3-1, originally presented by Dillon, Goldstein, and Schiffman [1978], which are not atypical of data analyzed in the social and business sciences. Three variables (home ownership, number of rooms in home, and length of residency) are dichotomous, four variables (location of previous home, marital status, head-of-household's occupation, and family life cycle) are qualitative–unordered categorical variables, and the remaining variables (head-of-household's education and family income) are categorical and ordered. If an analysis of the data were undertaken it would likely entail the following steps:

1. Numerical scores would be assigned to each level of each variable, with the exception of marital status, which would be coded as a dummy variable.
2. The total sample would be randomly split into two groups; one sample would be used to construct the function, and the second sample, generally called a *hold-out* sample, would be used for classification.

3. The function would be constructed in the first sample, and interest would be focused on its significance (i.e., a test of between-group differences) and the β coefficients.
4. The performance of the function would be investigated when applied to the hold-out sample.
5. If successful, the two samples would be combined, and the function would be recomputed using all of the available sample observations.

Whereas steps (2)–(5) seem intuitively reasonable and attempt to ensure some degree of validity, step (1) can produce some rather severe anomalies in the results. One of the first problems encountered with the assignment of numeric scores to the levels of the variables is the rather fictitious information introduced in the form of metrics on such qualitative constructs as occupational status or stage in family life cycle. Use of Fisher's LDF requires that the characteristics have been quantitatively measured. Clearly, in this case the assigned numbers do not strictly refer to the degree of a characteristic possessed by the individual; rather, they merely refer to the classes or categories of individuals themselves. Attempts to quantify qualitative data have generally employed an analytical technique, such as factor analysis, to transform the data into indices or scores before the function is constructed. Moreover, in cases where distance measures have been proposed requiring only nominal measurement, the properties of the resultant metric are not well defined.

Potentially more serious, the assignment of numerical values may, by its very nature, create either illusionary linearities in the data or, perhaps, introduce nonlinearities that cause variables to have opposite signs (the β coefficients in the fitted function), than what would be expected on the basis of prior knowledge. To see this, return to the variable descriptions given in Table 1.3-1 and the outline of the likely steps in analysis given on the previous page. As we indicated, the researcher is frequently concerned with the discriminant coefficients since they are used to facilitate model building. That is, if we standardize each coefficient by dividing by its respective standard deviation, then the resultant sign and magnitude of each coefficient provides a comparative measure of the discriminative ability of each variable. However, if the coefficients are to be useful we would at least hope that they exhibit a substantial degree of stability across separate samples.

To investigate this point further, we refer to the study by Dillon, Goldstein, and Schiffman [1978] which attempted to discriminate between heavy and light users of a major household service on the basis of the nine variables given in Table 1.3-1. The analysis followed steps (1)–(5). The

TABLE 1.3-1

SOME TYPICAL VARIABLES USED IN ANALYSIS IN MARKETING RESEARCH

VARIABLE	DESCRIPTIONS
Home ownership	Own, rent
Number of rooms in home	Less than five rooms, at least five rooms
Length of residency	Less than five years, at least five years
Location of previous home	Five categories: Within this same city or town Outside of this city but in same county Outside of this county but in same state In another state but within the U.S.A. Outside the U.S.A.
Marital status	Three categories: Married Single Widowed, separated, or divorced
Head-of-household's occupation	Ten categories: Professional, technical Manager, official, or proprietor Sales or clerical worker Craftsman or foreman Operators Laborer Service worker Housewife Student Retired
Head-of-household's education	Eight categories: Some grade school Grade school completed Some high school High school completed Some college College graduate Some graduate work Master's or doctorate degree
Family income	Eleven categories: Under $3,000 $ 3,001– 5,000 $13,001–15,000 $ 5,001– 7,000 $15,001–20,000 $ 7,001– 9,000 $20,001–25,000 $ 9,001–11,000 $25,001–30,000 $11,001–13,000 Over $30,000

8

TABLE 1.3-1*(continued)*

VARIABLE	DESCRIPTIONS
Stage in family life cycle	Six categories: Head of household less than 55 yr old, single (widowed, separated, or divorced), no children Head of household less than 55 yr old, married, no children Head of household less than 55 yr old, with children (none teenagers) Head of household less than 55 yr old, with children (at least one teenager) Head of household at least 55 yr old, employed Head of household at least 55 yr old, unemployed

Source. Dillon, Goldstein, and Schiffman [1978].

total sample was randomly divided into an analysis sample consisting of 70% of the entire sample (325 observations) and a 30% hold-out sample (130 observations). Since the hold-out sample was small relative to the analysis sample, the authors believed that there might be a good deal of variability simply due to the reduced number of observations. Therefore, they chose to perform a more conservative test in which the discriminant coefficients derived from the analysis sample were compared to those obtained from using all of the available sample observations (analysis plus hold-out samples).

Table 1.3-2 presents summary results from their study. The table gives the standardized discriminant coefficients along with their relative rankings based on absolute magnitudes for both the analysis and total samples. Although precise replication of the analysis sample results is unlikely, the table clearly indicates that substantial variation exists between the two sets of coefficients. In particular, note that there are sign reversals for three of the variables—home ownership, head of household's education, and family life cycle. Furthermore, contrary to prior knowledge, head-of-household's education enters with a negative sign. Although the two most important variables in the analysis sample—family income and location of previous home—demonstrate a fair degree of stability, the relative rankings of the other variables fluctuate considerably. The degree of instability is rather surprising considering that the analysis sample comprised 70% of the total sample. The authors concluded that it is reasonable to suspect that

TABLE 1.3-2

STANDARDIZED DISCRIMINANT COEFFICIENTS AND RANKINGS FOR THE
ANALYSIS AND TOTAL SAMPLES BASED ON THE VARIABLES GIVEN IN
TABLE 1.3-1

	70% ANALYSIS SAMPLE		TOTAL SAMPLE	
VARIABLE DESCRIPTION	COEFFICIENT[a]	RANK	COEFFICIENT[a]	RANK
Home ownership	0.04905	9	−0.07641	6
Number of rooms in home	0.07436	7	0.13707	4
Length of residence	0.01083	10	0.06821	7
Location of previous home	0.22086	2	0.28720	2
Marital status				
Married	−0.17920	3	−0.08932	5
Single	0.10293	6	0.20213	3
Head-of-household's occupation	0.07287	8	0.04148	9
Head-of-household's education	0.15154	4	−0.01143	10
Family income	0.86139	1	0.91399	1
Stage in family life cycle	−0.13757	5	0.04649	8

[a]A high discriminant score is associated with the heavy product user group.
Source. Dillon, Goldstein, and Schiffman [1978].

anomalies of this kind are caused at least in part by the assignment of numeric scores.

1.4 SOME CONCLUDING THOUGHTS

Although the common practice is clearly to treat qualitative data as if they were continuous and use Fisher's LDF, some authors have advocated the reverse procedure, which consists of converting continuous variables into discrete ones and using discrete discriminant techniques. This recommendation largely follows from the work done on error rates that is discussed in Chapter 3. In addition, another interesting situation can be noted. This case arises when the available data are composed of both binary and continuous variables. In particular, the recent work of Krzanowski [1975] using the location model is noteworthy, and we discuss this material in Chapter 5.

In conclusion, we have attempted to promote the idea that in cases where the data are qualitative it may be more natural to assume underlying discrete structures and proceed with discriminant techniques based on such characterizations. Hopefully, we have stimulated the reader's interest and curiosity enough to continue to Chapter 2, which presents a detailed discussion of the general class of discrete, and in particular multinomial, discriminant procedures.

CHAPTER 2

Discrete Classification Models

2.1 INTRODUCTION

In Chapter 1 we attempted to emphasize why researchers should take a closer look at the classification problem rather than using some rule derived from assuming that underlying distributions are multivariate normal. Although as discussed, Fisher's approach does not require normality, its justification in the likelihood ratio sense certainly does. Our concern in this chapter is in the formulation, discussion, and illustration of classification rules that can be derived when the available data are discrete and hence depart from anything resembling normality. In so doing the objective is to capture the distributional properties that generate the data and to utilize familiar optimality arguments in proposing procedures. In the sections that follow we again restrict our discussion to the two-group problem, but most of what is presented has immediate extension to the more general $K(>2)$ group case.

Now suppose a researcher has data that are generated by discrete random variables $X_1, X_2, \ldots, X_p$, each assuming at most a finite number of distinct values $s_1, s_2, \ldots, s_p$. The sample space $\mathcal{X}$ consists of $s = \Pi_j s_j$ points or states and for our purposes is assumed generated by the random vector $\mathbf{X} = (X_1, X_2, \ldots, X_p)$, having a multinomial distribution. We denote the class-conditional multinomial mass functions within two disjoint population groups, Π_1 and Π_2, by f_1 and f_2, respectively. If the two populations Π_1 and Π_2 are mixed with prior probabilities δ_1 and δ_2, then the unconditional density of $\mathbf{X}$ at $\mathbf{x}$ is given by

$$g(\mathbf{x}) = \delta_1 f_1(\mathbf{x}) + \delta_2 f_2(\mathbf{x}) = g_1(\mathbf{x}) + g_2(\mathbf{x}).$$

For the two-group problem a classification rule may be defined as an ordered partition $D = \langle D_1, D_2 \rangle$ of the sample space $\mathcal{X}$, where the rule D allocates an observation $\mathbf{x}$ to Π_i if and only if $\mathbf{x} \in D_i$, for $i = 1$ or 2. The

11

conditional probability of misclassification given $\mathbf{X} = \mathbf{x} \in D_i$ is

$$t(D|\mathbf{X} = \mathbf{x}) = g_j(\mathbf{x})/g(\mathbf{x}), \, i \neq j.$$

Averaging out over $\mathbf{x}$ yields the unconditional error rate

$$t(D) = E\{t(D|\mathbf{X})\} = \sum_{D_1} g_2(\mathbf{x}) + \sum_{D_2} g_1(\mathbf{x}). \qquad (2.1\text{-}1)$$

A rule is optimal if it minimizes the unconditional probability of misclassification. Stated more formally, if t is a function on the domain $\mathcal{D}$ of all rules, then D is optimal if $t(D) = t^* = \inf_{D' \in \mathcal{D}} t(D')$. A fundamental result attributed to Welch [1939] and extended by Hoel and Peterson [1949] shows that an optimal partition D^* must be characterized by $\mathbf{x} \in D_1^*$ if $g_1(\mathbf{x}) > g_2(\mathbf{x})$ or by $\mathbf{x} \in D_2^*$ if $g_1(\mathbf{x}) < g_2(\mathbf{x})$, and randomly assigned if $g_1(\mathbf{x}) = g_2(\mathbf{x})$. Hence

$$t^* = t(D^*) = \sum \min(g_1(\mathbf{x}), g_2(\mathbf{x})),$$

where the summation is over all states $\mathbf{x}$ belonging to $\mathcal{X}$. Furthermore, note that for any $\mathbf{x}$ it follows that $t^* \leqslant \frac{1}{2}$ since $\min(g_1(\mathbf{x}), g_2(\mathbf{x})) \leqslant \frac{1}{2} g(\mathbf{x})$, with t^* close to $\frac{1}{2}$, indicating that the two discriminant scores, $g_1(\mathbf{x})$ and $g_2(\mathbf{x})$, are similar.

This discussion points to various avenues that can be followed in attempting to approximate (and/or estimate) the optimal rule D^*. The most obvious way to proceed is to use the sample-based procedure, which results from estimating the discriminant scores assuming a full multinomial model and then plugging these estimates into (2.1-1). Another approach involves representing the multinomial densities in terms of fewer parameters (e.g., assuming $X_1, X_2, \dots, X_p$ are independent) and proceeding with the estimation of the induced discriminant scores. Still another approach takes the form of using model representations in terms of new parameters in describing the behavior of the state probabilities $f_i(\mathbf{x})$, $i = 1, 2$ for all $\mathbf{x}$ belonging to $\mathcal{X}$. All three approaches eventually lead to sample-based likelihood ratio rules that we hope will come "close" to parroting the partition D^*. In the next few sections we describe approaches of this nature and illustrate their use through examples.

2.2 SAMPLING AND THE FULL MULTINOMIAL MODEL

There are two alternative basic sampling situations that we use in defining sample-based rules: (a) N individuals are sampled from the mixed population or (b) independent samples of size n_1 and n_2 are sampled respectively from Π_1 and Π_2 with prior probabilities usually specified. In the former case the number of individuals from Π_i with $\mathbf{X} = \mathbf{x}$ is a binomial random variable $N_i(\mathbf{x})$ with expected value $N\delta_i f_i(\mathbf{x})$, for $i=1$, 2. In addition, the total number of sampled individuals from the two populations $N_1 = \Sigma N_1(\mathbf{x})$ and $N_2 = \Sigma N_2(\mathbf{x})$ are binomial random variables with $N = N_1 + N_2$. Intuitive estimates for prior probabilities are given by $\hat{\delta}_i = (N_i/N)$, and the usual nonparametric estimates of the class-conditional densities or state probabilities, by $\hat{f}_i(\mathbf{x}) = N_i(\mathbf{x})/N_i$, $i=1$, 2. Hence it follows that the estimated discriminant scores for this case are given by $\hat{g}_i(\mathbf{x}) = N_i/N \cdot N_i(\mathbf{x})/N_i = N_i(\mathbf{x})/N$, for $i=1$, 2.

In the case of independent random samples, prior probabilities are usually not estimated but are specified. Further, $n_1(\mathbf{x})(n_2(\mathbf{x}))$ are random variables defined as the number of individuals from $\Pi_1(\Pi_2)$ with $\mathbf{X}=\mathbf{x}$. In this case $En_i(\mathbf{x}) = n_i f_i^*(\mathbf{x})$, $i=1$, 2, where the notation $f_i^*(\mathbf{x})$ is used to denote the density at $\mathbf{x}$ or the state probability defined by $\mathbf{x}$ from Π_i, as opposed to the class-conditional density $f_i(\mathbf{x})$. A nonparametric estimate of the density at $\mathbf{x}$ is given by $n_1(\mathbf{x})/n_1$ for $f_1^*(\mathbf{x})$, and if $\delta^*(>0)$ is the specified prior probability associated with Π_1, then the discriminant score for Π_1 is $\delta^* n_1(\mathbf{x})/n_1$.

An intuitive sample-based partition $\hat{D}$ of the sample space $\mathcal{X}$ is defined by $\hat{D} = \langle \hat{D}_1, \hat{D}_2 \rangle$ with $\mathbf{x} \in \hat{D}_1$ if $\hat{g}_1(\mathbf{x}) > \hat{g}_2(\mathbf{x})$, $\mathbf{x} \in \hat{D}_2$ if $\hat{g}_1(\mathbf{x}) < \hat{g}_2(\mathbf{x})$ and is randomly assigned if $\hat{g}_1(\mathbf{x}) = \hat{g}_2(\mathbf{x})$. Thus when sampling from the mixed population, the sample-based classification rule becomes: $\mathbf{x} \in \hat{D}_1$ if $N_1(\mathbf{x}) > N_2(\mathbf{x})$ or $\mathbf{x} \in \hat{D}_2$ if $N_2(\mathbf{x}) > N_1(\mathbf{x})$. A point with $N_1(\mathbf{x}) = N_2(\mathbf{x}) \geqslant 0$ will be assigned to $\hat{D}_1$ or to $\hat{D}_2$ with equal probability. In the case of independent samples a similar idea prevails, except that: $\mathbf{x} \in \hat{D}_1$ if $\delta^* n_1(\mathbf{x})/n_1 > (1-\delta^*)n_2(\mathbf{x})/n_2$, $\mathbf{x} \in \hat{D}_2$ if $\delta^* n_1(\mathbf{x})/n_1 < (1-\delta^*)n_2(\mathbf{x})/n_2$, and randomly assigned if otherwise.

Both sample-based rules have obvious appeal due to their intuitiveness and ease of use. We have more to say concerning asymptotic properties of these rules in the next chapter, where the problem of error rates is discussed. Note also that in the examples that follow we refer to the error rates as simply the *misclassification probabilities*.

There are, however, certain observations that are immediately apparent and point to potential difficulties in applying the so-called full multinomial rule. Perhaps of overriding concern, is the problem of state proliferation made especially troublesome in practice by the availability of relatively small sample sizes. For example, five variables, each assuming only three distinct values, generate 243 states. Therefore, to guarantee that all states are nonempty, we would need to sample at least 243 observations from each population. Obviously, a large number of observations relative to the number of variables is required if sufficient data in each state are to be available for the estimation of state probabilities.

Aside from the problem of zeros in states or the potential of few observations on which to base the estimation of state probabilities is the issue that for a given state a zero from Π_1 might mean something entirely different from a zero coming from Π_2. Moreover, especially if samples of disproportionate size are available, error rates generated by the somewhat forced allocation caused by a zero in, say, state j from Π_1, and a nonzero in state j from Π_2 can be potentially misleading. It is because of these difficulties that some researchers have been reluctant to apply full multinomial procedures in situations where the data set suffers from severe sparseness. We have more to say about the problem of disproportionate sample sizes in Section 2.4.

2.2.1 Variations of the Full Multinomial Model

The two variants of the full multinomial procedures that have received some attention in the literature assume that the data are dichotomous. In particular, suppose that $X_j = 0$ or 1 for $j = 1, 2, \ldots, p$. It follows that the induced multinomial distributions have $s = 2^p$ states, with $2^p - 1$ independent parameters in each population. As a means of reducing the number of parameters needed to be estimated, one procedure assumes that the variables are all independent. This assumption, although probably unrealistic in many applications, reduces the number of parameters in each population from $2^p - 1$ to p, and we refer to this procedure as the *first-order independence model*.

Hence for a given vector of observations $\mathbf{x} = (x_1, x_2, \ldots, x_p)$, $x_j = 0$ or 1, $j = 1, 2, \ldots, p$, the probability $P\{\mathbf{X} = \mathbf{x} | \Pi_i\}$, $i = 1, 2$, is given by

$$\prod_{j=1}^{p} \left(P\{X_j = x_j | \Pi_i\} \right)^{x_j} \cdot \left(1 - P\{X_j = x_j | \Pi_i\} \right)^{1 - x_j}. \qquad (2.2\text{-}1)$$

Further, assuming independent samples of size n_i from Π_i, $i = 1, 2$, unbiased estimates for $P\{X_j = x_j | \Pi_i\}$ are

$$\hat{P}\{X_j = x_j | \Pi_i\} = \sum_{s_j} \frac{n_i(\mathbf{x})}{n_i} \qquad (2.2\text{-}2)$$

where s_j is the set of responses $\mathbf{x}$ where $X_j = x_j$. The sample-based classification rule that now evolves is given by

Classify x in Π_1 if $\delta^* \prod_{j=1}^{p} \sum_{s_j} \frac{n_1(\mathbf{x})}{n_1} > (1 - \delta^*) \prod_{j=1}^{p} \sum_{s_j} \frac{n_2(\mathbf{x})}{n_2}$

Classify x in Π_2 if $\delta^* \prod_{j=1}^{p} \sum_{s_j} \frac{n_1(\mathbf{x})}{n_1} < (1 - \delta^*) \prod_{j=1}^{p} \sum_{s_j} \frac{n_2(\mathbf{x})}{n_2}$

Randomly allocate if $\delta^* \prod_{j=1}^{p} \sum_{s_j} \frac{n_1(\mathbf{x})}{n_1} = (1 - \delta^*) \prod_{j=1}^{p} \sum_{s_j} \frac{n_2(\mathbf{x})}{n_2}$ $\qquad (2.2\text{-}3)$

This classification rule clearly does not suffer, to the same extent, from potential difficulties of a sparse data set. However, its usefulness is suspect because of the assumption of independence.

The second variation, originally proposed by Hills [1967] as a method for reducing the sampling error of the sample-based full multinomial rule, is the ad hoc nearest neighbor of order r rule. The idea of Hill's method is that when using a sample-based likelihood ratio procedure for classifying a particular response vector x, all responses differing from x in no more than r components are incorporated into the rule. In particular, for a given response vector x, let $T_j = \{y_j | (x - y_j)' (x - y_j) \leqslant r\}$. Note that T_j is merely the set of responses $\{y_j\}$ having the property that each of its elements differs in no more than r components from x.

Assuming independent samples of size n_i from Π_i, $i = 1, 2$, the nearest neighbor of order r rule is given by

$$\text{Classify } x \text{ into } \Pi_1 \text{ if } \delta^* \sum_{T_j} \frac{n_1(y_j)}{n_1} > (1 - \delta^*) \sum_{T_j} \frac{n_2(y_j)}{n_2}$$

$$\text{Classify } x \text{ into } \Pi_2 \text{ if } \delta^* \sum_{T_j} \frac{n_1(y_j)}{n_1} < (1 - \delta^*) \sum_{T_j} \frac{n_2(y_j)}{n_2}$$

$$\text{Randomly allocate if } \delta^* \sum_{T_j} \frac{n_1(y_j)}{n_1} = (1 - \delta^*) \sum_{T_j} \frac{n_2(y_j)}{n_2} \qquad (2.2\text{-}4)$$

Note that when $r = 0$, (2.2-4) is equivalent to the full multinomial rule.

Example 2.2-1   Information and Store Choice.   The data for our first example were collected and reported in a somewhat abridged form in the *Journal of Advertising Research* (Dash, Schiffman, and Berenson [1977]). These authors examined whether shoppers of two markedly different retail establishments—a full-line department store (Π_1) or an audio equipment specialty store (Π_2)—differed in terms of their information-related predispositions and information-seeking activities. Questionnaires were completed by 464 respondents of which 157 were identified as shoppers of the full-line department store, and the remaining 267 were customers of the specialty store.

Table 2.2-1 presents the observed frequency distributions for the two shopper groups with respect to four dichotomous variables. The variables are:

Variable 1:   Information seeking

> 1   if the individual sought information from friends and/or neighbors before purchase.
>
> 0   otherwise.

Variable 2:   Information transmitting

    1   if the individual had recently been asked for an opinion about buying any audio product.

    0   otherwise.

Variable 3:   Prior shopping experience

    1   if the individual had shopped in any stores for audio equipment before making a decision.

    0   otherwise.

Variable 4:   Catalog experience

    1   if the individual had sought information from manufacturers' catalogs before purchase.

    0   otherwise.

TABLE 2.2-1

OBSERVED FREQUENCY DISTRIBUTIONS FOR FULL-LINE DEPARTMENT
AND SPECIALTY STORE SHOPPERS[a]

STATE $(x_1x_2x_3x_4)$	FULL-LINE DEPARTMENT STORE (Π_1)		SPECIALTY STORE (Π_2)	
	OBSERVED	RELATIVE	OBSERVED	RELATIVE
1 1 1 1	5	0.033	86	0.333
1 1 1 0	2	0.013	22	0.085
1 1 0 1	15	0.098	23	0.089
1 1 0 0	4	0.026	11	0.043
1 0 1 1	3	0.019	3	0.012
1 0 1 0	3	0.019	4	0.016
1 0 0 1	3	0.019	4	0.016
1 0 0 0	5	0.033	3	0.012
0 1 1 1	14	0.091	33	0.128
0 1 1 0	8	0.052	6	0.023
0 1 0 1	26	0.169	30	0.116
0 1 0 0	12	0.078	5	0.019
0 0 1 1	2	0.013	8	0.031
0 0 1 0	3	0.019	6	0.023
0 0 0 1	32	0.208	8	0.031
0 0 0 0	17	0.110	6	0.023
	154	1.000	258	1.000

[a]The analysis is based on 412 observations instead of 424 since 12 observations were omitted due to nonresponse.

The data are used to demonstrate the full multinomial, first-order independence and nearest-neighbor rules discussed in this section. Note that in the case of the nearest-neighbor rule we consider only the rule determined by near neighbors of order $r=1$.

To make the nearest-neighbor rule operational requires that the set of near neighbors for each set be specified. For example, in our case the near neighbors of order $r=1$ of $\mathbf{x}=(1111)$ are given by

$$T_{1111}=0111,1011,1101,1110$$

Hence from the data of Table 2.2-1 we see that the new state frequencies for $\mathbf{x}=(1111)$ are $(5+3+15+2+14)=39$ in Π_1 and $(86+3+23+22+33)=167$ in Π_2. A similar computation is required for each of the states, and the recomputed set of frequency distributions is given in Table 2.2-2.

To evaluate each of the three classification rules we initially assume equal prior probabilities, that is, $\delta^*=0.5$. For each of the procedures we form the ratio of likelihoods and classify $\mathbf{X}=\mathbf{x}$ into Π_1 if $\hat{P}\{\mathbf{X}=\mathbf{x}|\Pi_2\}/\hat{P}\{\mathbf{X}=\mathbf{x}|\Pi_1\}<\delta^*/(1-\delta^*)$, where $\hat{P}\{\mathbf{X}=\mathbf{x}|\Pi_i\},i=1,2$, is the appropriate estimate of the full multinomial, first-order independence or nearest-neighbor procedure. Applying each of the rules yields the following

TABLE 2.2-2

FREQUENCY DISTRIBUTIONS FOR NEAR NEIGHBORS OF ORDER $r=1$ RULE

STATE $(x_1x_2x_3x_4)$	FULL-LINE DEPARTMENT STORE (Π_1)	SPECIALTY STORE (Π_2)
1 1 1 1	39	167
1 1 1 0	22	129
1 1 0 1	53	154
1 1 0 0	38	64
1 0 1 1	16	105
1 0 1 0	16	38
1 0 0 1	58	41
1 0 0 0	32	28
0 1 1 1	55	163
0 1 1 0	39	72
0 1 0 1	99	99
0 1 0 0	67	58
0 0 1 1	54	58
0 0 1 0	33	30
0 0 0 1	80	56
0 0 0 0	69	28

Discrete Classification Models

misclassification probabilities: (a) 0.2757 for the full multinomial, (b) 0.3110 for the first-order independence, and (c) 0.3110 for the nearest neighbor.

To clearly see how the three rules lead to different assignments, consider Table 2.2-3, which shows the logarithm of the likelihood ratio and the corresponding group designation for each x. Note that the first-order independence and near-neighbor methods yield equivalent allocation rules in each of the states and, therefore, we obtain identical error rates with their use. On the other hand, the full multinomial procedure yields an allocation rule that differs from the other procedures in states:

$$1101, 1011, 1010$$

$$0110, 0011, 0010.$$

Heretofore we have assumed equal prior probabilities. In general, however, substantially different results are obtained if the prior probabilities are estimated and set equal to the sample-group frequencies rather than being assumed equal, that is, $\hat{\delta}_i = (N_i/N)$, $i = 1, 2$. If the estimated prior probabilities are used, then the misclassification probabilities reevaluate

TABLE 2.2-3

VALUES OF $\log[P\{\mathbf{X}=\mathbf{x}|\Pi_2\}/P\{\mathbf{X}=\mathbf{x}|\Pi_1\}]$ FOR THE FULL MULTINOMIAL, FIRST-ORDER INDEPENDENCE AND NEAREST-NEIGHBOR RULES

STATE $(x_1x_2x_3x_4)$	FULL MULTINOMIAL		FIRST-ORDER INDEPENDENCE		NEAREST NEIGHBOR	
	$\log[\cdot]$	ALLOCATION	$\log[\cdot]$	ALLOCATION	$\log[\cdot]$	ALLOCATION
1 1 1 1	2.329	Π_2	2.321	Π_2	0.938	Π_2
1 1 1 0	1.882	Π_2	1.807	Π_2	1.253	Π_2
1 1 0 1	-0.089	Π_1	0.649	Π_2	0.551	Π_2
1 1 0 0	0.496	Π_2	0.135	Π_2	0.005	Π_2
1 0 1 1	-0.516	Π_1	0.918	Π_2	1.365	Π_2
1 0 1 0	-0.228	Π_1	0.404	Π_2	0.349	Π_2
1 0 0 1	-0.228	Π_1	-0.753	Π_1	-0.863	Π_1
1 0 0 0	-1.027	Π_1	-1.267	Π_1	-0.649	Π_1
0 1 1 1	0.341	Π_2	0.848	Π_2	0.570	Π_2
0 1 1 0	-0.804	Π_1	0.335	Π_2	0.097	Π_2
0 1 0 1	-0.373	Π_1	-0.823	Π_1	-0.516	Π_1
0 1 0 0	-1.391	Π_1	-1.337	Π_1	-0.660	Π_1
0 0 1 1	0.870	Π_2	-0.554	Π_1	-0.445	Π_1
0 0 1 0	0.177	Π_2	-1.068	Π_1	-0.611	Π_1
0 0 0 1	-1.902	Π_1	-2.226	Π_1	-0.873	Π_1
0 0 0 0	-1.557	Π_1	-2.739	Π_1	-1.418	Π_1

to: (a) 0.2621 for the full multinomial, (b) 0.3010 for the first-order independence, and (c) 0.2803 for the nearest neighbor. Note that using the estimated prior probabilities increases the classification efficiency for all the methods. In general, this is true since the population with the larger sample size tends to be assigned more of the sample observations than it would otherwise. However, this is not an unmixed blessing since the use of the sample group frequencies as estimates of the prior probabilities often prohibits a meaningful analysis of the smaller population and, in addition, may lead to results that reflect variations in sample sizes rather than true differences between the populations.

2.3   MODEL REPRESENTATIONS FOR MULTIVARIATE DICHOTOMOUS RESPONSES

Several authors, interested in making inferences assuming as a starting condition multivariate dichotomous structures, have developed representations of underlying joint densities in terms of various reparametrization schemes. Although this section is not meant to be a survey of all such efforts, we discuss most of the available representations with respect to their implementation in developing classification rules. As a general statement, the number of new parameters utilized in all such reparametrizations is essentially equal to the number of parameters needed to characterize the full multinomial, that is, $2^p - 1$. Hence, if full representations are used, the same prevailing problem of a large number of parameters coupled with a relatively small data set may recur. However, in many instances the value of a given state probability when expressed in such forms is mostly captured when only a few parameters are used. Translated into the classification problem, this means that error rates may often be comparable to using full structures when reduced models are used, while at the same time having the capacity to deal with the problem of sparseness.

In the remaining sections discussing reparametrizations we use f to represent a general multivariate dichotomous density, whereas f_1 and f_2 are the corresponding densities associated with populations Π_1 and Π_2, respectively.

2.3.1   The Bahadur Model

The first model discussed is attributed to Bahadur [1961]. We assume that $\mathbf{X} = (X_1, X_2, \ldots, X_p)$, where X_j is a Bernoulli random variable with $\theta_j = P\{X_j = 1\} = E(X_j)$ and $1 - \theta_j = P\{X_j = 0\}, j = 1, 2, \ldots, p$. Now set

$$Z_j = \frac{X_j - \theta_j}{\sqrt{\theta_j(1 - \theta_j)}}$$

and define

$$\rho_{jk} = E(Z_j Z_k)$$

$$\rho_{jkl} = E(Z_j Z_k Z_l)$$

$$\vdots$$

$$\rho_{jk\dots,p} = E(Z_j Z_k \cdots Z_p) \tag{2.3-1}$$

Let $p_{[1]}(x_1, x_2, \dots, x_p)$ denote the joint-probability distribution of the x_j when the x_j values: (a) are independently distributed and (b) have the same marginal distributions as under the given distribution $f(x_1, x_2, \dots, x_p)$. Then with

$$p_{[1]}(x_1, x_2, \dots, x_p) = \prod_{j=1}^{p} \theta_j^{x_j} (1 - \theta_j)^{1 - x_j} \tag{2.3-2}$$

Bahadur has shown that

$$f(\mathbf{x}) = p(\mathbf{x}) p_{[1]}(\mathbf{x})$$

where

$$p(\mathbf{x}) = 1 + \sum_{j<k} \rho_{jk} z_j z_k + \sum_{j<k<l} \rho_{jkl} z_{jkl} + \cdots + \rho_{12\dots p} z_1 z_2 \cdots z_p. \tag{2.3-3}$$

In many applications it makes good sense to assume that higher-order correlations are zero, hence reducing the number of parameters required for estimation. Of course, if all parameters are known and used, then the likelihood ratio rule will generate smaller misclassification probability than if only a subset is used. However, if parameters need to be estimated, then it is quite possible, because of the potential instability of the estimated parameters, to obtain smaller sample error rates for the reduced models than for saturated models. Solomon [1961] used a Bahadur representation in discussing a particular classification problem, whereas Moore [1973] used it in a Monte Carlo study to compare various procedures.

　　Whichever representation is decided on, the classification procedure used is the simple likelihood ratio with estimates replacing the θ_j and correlation parameters in each population. In the case where all correlation terms beyond order 1 are assumed zero, the mass function is approximated and estimated by

$$\hat{f}(\mathbf{x}) = \prod_{j=1}^{p} \hat{\theta}_j^{x_j} \left(1 - \hat{\theta}_j\right)^{1-x_j} \left\{ 1 + \sum_{j<k} \hat{\rho}_{jk} \hat{z}_j \hat{z}_k \right\} \qquad (2.3\text{-}4)$$

where

$$\hat{\theta}_j = \sum_{s_j} \frac{n(\mathbf{x})}{n},$$

$$\hat{\rho}_{jk} = \frac{\sum\limits_{s_{jk}} n(\mathbf{x})/n - \hat{\theta}_j \hat{\theta}_k}{\sqrt{\hat{\theta}_j\left(1 - \hat{\theta}_j\right)\hat{\theta}_k\left(1 - \hat{\theta}_k\right)}}$$

Note that s_j is the set of all patterns $\mathbf{x}$ with $x_j = 1$, whereas s_{jk} is the set of $\mathbf{x}$ with $x_j = 1$ and $x_k = 1$. To distinguish parameters under Π_1 as opposed to those from Π_2 it is often customary to write θ_{ij} for $P\{X_j = 1|\Pi_i\}$, $i = 1, 2$, and $\rho_i(jk)$ for the corresponding correlation terms.

Example 2.3-1   Solomon Data—Attitudes Toward Science. Solomon [1961] reported the results of an "Attitude Toward Science" study conducted on a sample of high school students. A total of 2982 responses on four variables were received, of which (for our purposes we may assume) 1491 were identified from a high IQ group, and the remaining 1491 were from a lower IQ group. Whereas originally respondents were allowed five choices along an "agree–disagree" scale, their responses were collapsed and recorded as 1 or 0 for "agree" or "disagree." The four variables are:

X_1: The development of new ideas is the scientist's greatest source of satisfaction.

X_2: Scientists and engineers should be eliminated from the military.

X_3: The scientist will make his maximum contribution to society when he has freedom to work on problems that interest him.

X_4: The monetary compensation of a Nobel prize winner in physics should be at least equal to that given popular entertainers.

The observed frequency distributions for the four items are given in Table 2.3-1.

Discrete Classification Models

TABLE 2.3-1

OBSERVED FREQUENCY DISTRIBUTIONS FOR
SOLOMON DATA: ATTITUDES TOWARD SCIENCE

STATE $(x_1x_2x_3x_4)$	LOW IQ (Π_1)		HIGH IQ (Π_2)	
	OBSERVED	RELATIVE	OBSERVED	RELATIVE
1 1 1 1	62	0.042	122	0.082
1 1 1 0	70	0.047	68	0.046
1 1 0 1	31	0.021	33	0.022
1 1 0 0	41	0.027	25	0.017
1 0 1 1	283	0.190	329	0.221
1 0 1 0	253	0.170	247	0.166
1 0 0 1	200	0.134	172	0.115
1 0 0 0	305	0.205	217	0.146
0 1 1 1	14	0.009	20	0.013
0 1 1 0	11	0.007	10	0.007
0 1 0 1	11	0.007	11	0.007
0 1 0 0	14	0.009	9	0.006
0 0 1 1	31	0.021	56	0.037
0 0 1 0	46	0.031	55	0.037
0 0 0 1	37	0.025	64	0.043
0 0 0 0	82	0.055	53	0.035
	1491	1.000	1491	1.000

Using the Bahadur reparametrization the parameters evaluate to

IQ	$\hat{\rho}_{12}$	$\hat{\rho}_{13}$	$\hat{\rho}_{14}$	$\hat{\rho}_{23}$	$\hat{\rho}_{24}$	$\hat{\rho}_{34}$	$\hat{\rho}_{123}$	$\hat{\rho}_{124}$	$\hat{\rho}_{134}$	$\hat{\rho}_{234}$	$\hat{\rho}_{1234}$
High	0.003	0.143	0.111	0.180	0.020	0.140	-0.010	-0.064	-0.041	-0.033	0.003
Low	0.049	0.144	0.043	0.155	0.096	0.125	-0.006	-0.012	0.002	0.002	0.002

and

IQ	$\hat{\theta}_1$	$\hat{\theta}_2$	$\hat{\theta}_3$	$\hat{\theta}_4$
High	0.821	0.159	0.505	0.436
Low	0.801	0.189	0.599	0.530

The question now is which representation to use. For example, if a saturated model is selected, the error rate evaluates to 0.4413. However, as we have indicated, the crucial question is how well reduced models can perform.

Two reduced representations most often examined evolve from assuming: (a) all correlation terms zero and (b) correlations beyond those of the first-order zero. It is common practice to refer to the former specification as the *first-order model* since only the marginal distributions of each X_j are used in approximating the state probabilities, whereas the latter case is referred to as the *second-order model*. Applying these two reduced models to the data of Table 2.3-1 yields the following misclassification probabilities: (a) 0.4423 for the first-order model and (b) 0.4420 for the second-order model. Although in this case relatively poor performance results no matter which representation is selected, we do see that use of either of the reduced models yields essentially the same error rate as the saturated model.

2.3.2   Loglinear and Logit Methods

The model that probably has received greatest attention in the literature expresses the logarithm of state probabilities as a linear combination of main effects and interactions. In particular, suppose that

$$\log f(\mathbf{x}) = \alpha + \sum_{j=1}^{p} (-1)^{x_j} \alpha_j + \sum_{j<k} (-1)^{x_j+x_k} \alpha_{jk} + \cdots$$

$$\vdots$$

$$+ (-1)^{x_1+x_2+\cdots+x_p} \alpha_{12\cdots p} \qquad (2.3\text{-}5)$$

where α is an overall effect, α_j is the main effect due to X_j, α_{jk} is the interaction effect due to X_j and X_k, and so on. By design it is assumed that each state defined by its vector $\mathbf{x} = (x_1, x_2, \ldots, x_p)$ is such that $f(\mathbf{x}) > 0$.

Suppose it is assumed that $\log[f_1(\mathbf{x})/f_2(\mathbf{x})] = 2\boldsymbol{\beta}'\tilde{x}$, where $\boldsymbol{\beta}' = (\beta_0, \beta_1, \ldots, \beta_p)$, and $\tilde{x} = (1, \mathbf{x})$. Under this assumption it is readily demonstrated that, assuming all second- and higher-order interaction terms are zero,

$$\beta_k = 2\alpha_{kp} \, (k = 1, 2, \ldots, p-1)$$

$$\beta_0 = -\alpha_p - \sum_{k=1}^{p-1} \frac{\beta_k}{2}. \qquad (2.3\text{-}6)$$

Berkson [1955] has proposed that the vector of parameters $\boldsymbol{\beta}' = (\beta_0, \beta_1, \ldots, \beta_p)$ be estimated using minimum logit χ^2 estimates. Based on random samples of size n_1 and n_2 from Π_1 and Π_2, the vector $\hat{\boldsymbol{\beta}}' =$

$(\hat{\beta}_0, \hat{\beta}_1, \ldots, \hat{\beta}_p)$ that minimizes

$$\sum_x \frac{n_1(x)n_2(x)}{n(x)} \left[\log \frac{n_1(x)}{n_2(x)} - 2\beta'\tilde{x} \right]^2, \tag{2.3-7}$$

where $n(x)$ is $n_1(x) + n_2(x)$, is called a *minimum logit estimate*. In addition, Berkson recommends that when $n_i(x) = 0, i = 1, 2$, the value $\frac{1}{2}$ be substituted and when $n_i(x) = n(x)$, the value $n(x) - \frac{1}{2}$ be substituted.

Note that if $q_x = P\{\Pi = \Pi_1 | X = x\}$, then with $P\{\Pi = \Pi_1\} = \frac{1}{2}$

$$\log \frac{f_1(x)}{f_2(x)} = \log \left[\frac{q_x}{(1 - q_x)} \right] \tag{2.3-8}$$

the logit of q_x. Therefore, under the assumption that

$$\log \frac{f_1(x)}{f_2(x)} = 2\beta'\tilde{x} \tag{2.3-9}$$

it follows that

$$q_x = \frac{1}{1 + \exp - 2\beta'\tilde{x}} \tag{2.3-10}$$

Cox [1970] and Day and Kerridge [1967] discuss logistic discrimination assuming sampling from the mixture of the populations. The method has been extended by Anderson [1972] to more than two populations and to the sampling scheme where independent samples are selected from Π_1 and Π_2. Of course, the success of the logistic approach to discrimination depends on the extent to which the equations for the probabilities (2.3-8–10) are satisfied. However, the relations hold exactly in many situations, including: (a) multivariate normal populations with equal dispersion matrices, (b) multivariate independent dichotomous structures, and (c) multivariate dichotomous variables following a loglinear model with equal second- and higher-order effects.

Maximum likelihood estimates may be obtained by maximizing with respect to β the likelihood function

$$L = \prod_x q_x^{n_1(x)}(1 - q_x)^{n_2(x)} \tag{2.3-11}$$

where $q_x = 1/(1 + \exp - 2\beta'\tilde{x})$. The solution $\hat{\beta}'$ must satisfy the equations

$$n_{1k} = \sum_{x} x_k n(x) \left[\frac{1}{1 + \exp - 2\beta'\tilde{x}} \right] \qquad (2.3\text{-}12)$$

$k = 1, 2, \ldots, p$, where n_{1k} is the number of observations from Π_1 with $X_k = 1$ and $n_1(x) + n_2(x) = n(x)$.

The equations cannot be solved explicitly, so that an iteration procedure like the Newton–Raphson procedure is used. In practice, this has given very good convergence using zeros as starting values for all the coefficients. For a much more in-depth discussion of procedures and problems with the iteration scheme, the interested reader is directed to Cox [1970] and Anderson [1972].

It should be noted that other authors (Truett, Cornfield, and Kannel [1967]) have suggested an alternative approach to the estimation of (2.3-10), namely, the use of linear discriminant-function theory to estimate the $\hat{\beta}$ coefficients. For this approach (2.3-10) arises from the assumption of normality of the x or, at least, the assumption of normality for the β'. Although *a priori* the maximum likelihood approach is to be preferred, it is important to note that the discriminant-function approach is noniterative and computationally simpler than the maximum likelihood solution. In addition, under the assumption of normality and equal covariance matrices, the discriminant-function approach yields estimates that are unconditional maximum likelihood estimates, whereas the maximum likelihood solution yields (maximum likelihood) estimates conditional on the x. However, as Halperin, Blackwelder, and Verter [1971] have shown, although the maximum likelihood approach will secure a good fit if the model holds, it is theoretically possible for the discriminant-function approach to give a very poor fit even if the model is appropriate.

As was demonstrated in Chapter 1, for certain population structures the log likelihood ratio formed is not monotone with respect to the number of positive X_j. It is reasonable, therefore, to suspect that assuming as a starting condition the model

$$\log \frac{f_1(x)}{f_2(x)} = 2\beta'\tilde{x}$$

may lead (for certain problems) to large error probabilities.

Another approach to utilizing loglinear models as representations of state probabilities is to view the discrimination problem within the context of multidimensional contingency-table analysis. Sampling from the mixed population would correspond to generating a table through a multinomial sampling scheme, whereas fixed independent samples from Π_1 and Π_2 would be the equivalent of product-multinomial sampling. As an illustration of these ideas, suppose we consider a three-variable (X_1, X_2, X_3) problem which under one of two sampling schemes generates a table of dimension $2 \times J \times K$. Note that if $J = K = 2$ we would be in the multivariate dichotomous case being considered in this section. In general, however, J and K can be any positive integer greater than or equal to 2. We view the first variable as one identifying the two groups of interest, that is $X_1 = 1 \Leftrightarrow \Pi_1$ while $X_1 = 0 \Leftrightarrow \Pi_2$.

If we denote the state probability corresponding to cell (i, j, k) by p_{ijk}, then a hierarchical loglinear model for p_{ijk} would assume the form

$$\ln p_{ijk} = U + U_1(i) + U_2(j) + U_3(k) + U_{12}(ij) + U_{13(ik)} + U_{23(jk)} + U_{123(ijk)}$$

where

$$\Sigma_i U_1(i) = \Sigma_j U_2(j) = \Sigma_k U_3(k) = 0$$

$$\Sigma_i U_{12(ij)} = \Sigma_j U_{12}(ij) = \Sigma_i U_{13}(ik) = \Sigma_k U_{12}(ik) = \Sigma_j U_{23}(jk)$$

$$= \Sigma_k U_{23}(jk) = \Sigma_i U_{123}(ijk) = \Sigma_j U_{123}(ijk) = \Sigma_k U_{123}(ijk) = 0$$

Where all parameters (U-terms) are utilized in the representation for $\ln p_{ijk}$, we have a saturated model and maximum likelihood estimates of the U-terms when combined will be equal to the usual frequency estimates of $\ln p_{ijk}$. In general, however, one wants to fit an unsaturated model (i.e., one with fewer parameters) by determining maximum likelihood estimates for only a subset of the U-parameters. Without going into any detail on how this can be accomplished (e.g., by using the iterative proportional-fitting algorithm), the reader is advised to consult Bishop, Fienberg, and Holland [1975] or Fienberg [1977]. The big advantage to this approach is that we do not *a priori* eliminate particular interaction terms but utilize goodness-of-fit statistics as a tool of model building.

Given an observation characterized by $X_2 = j$ and $X_3 = k$, the sample-based rule (using equal priors) would be: classify into $\Pi_1(\Pi_2)$ if

$$\ln \frac{\hat{p}_{1jk}}{\hat{p}_{2jk}} > (<)0.$$

If the sampling scheme is multinomial then $\hat{p}_{1jk}$ and $\hat{p}_{2jk}$ are maximum likelihood estimates found by fitting an appropriate unsaturated hierarchical model. On the other hand, if independent samples are selected from Π_1 and Π_2 then the left-hand side of the optimal rule, namely, $\ln (\hat{p}_{1jk}/\hat{p}_{2jk})$, may be interpreted as a logit model with certain interaction terms included due to constraints on fixed marginal totals. In general, the two sampling schemes will result in distinct classification rules.

Although this approach has at this writing not appeared in the literature, it appears that it could have great impact on the discrimination problem. In particular, as long as given margins of the table contain positive frequencies there is a good chance that the state-sparseness problem will significantly be helped. In addition, if some of the categorical variables have inherent ordering then this information can be incorporated into the model-building process and have impact on the allocation rule. Approaches to ordered categorical data can be found in Haberman [1974] and Simon [1974].

Example 2.3-2   A Numerical Illustration.   To illustrate the computation of the various estimates, we show an example originally presented by Gilbert [1968] for the simplest case, $p = 2$.

Following (2.3-5) and assuming all second- and higher-order interactions to be zero, the parameters of the example population are taken to be

α_1	α_2	α_3	α_{12}	α_{13}	α_{23}
-0.2	-0.8	0.0	0.4	0.2	0.5

where α is determined so that $f(\mathbf{x})$ sum to unity. By (2.3-6), straightforward calculations show that

β_0	β_1	β_2
-0.7	0.4	1.0

The $f(\mathbf{x})$ are then as follows:

$\mathbf{x}$	(11)	(10)	(01)	(00)
$f_1(\mathbf{x})$	0.5488	0.0183	0.1108	0.0183
$f_2(\mathbf{x})$	0.1353	0.0337	0.0608	0.0743

Suppose a random sample of size 50 is now drawn from a population characterized by these parameters with observed frequency distributions in Π_1 and Π_2 given by

$\mathbf{x}$	(11)	(10)	(01)	(00)
Π_1	29	1	5	0
Π_2	6	3	4	2

TABLE 2.3-2

ESTIMATES OF $\log[f_1(\mathbf{x})/f_2(\mathbf{x})]$

STATE $(x_1 x_2)$	$f(\mathbf{x})$	MAXIMUM LIKELIHOOD	MINIMUM x^2
1 1	.684	1.60	1.48
1 0	.052	-1.29	$-.72$
0 1	.172	.16	.35
0 0	.093	-2.73	-1.86
Correlation coefficient r		.99787	.99711

Source. Adopted from Gilbert [1968].

As was indicated, the estimates for the maximum-likelihood approach are obtained by maximizing with respect to $\boldsymbol{\beta}$ the likelihood function

$$L = \prod_{\mathbf{x}} q_{\mathbf{x}}^{n_1(\mathbf{x})}(1 - q_{\mathbf{x}})^{n_2(\mathbf{x})}$$

where $q_{\mathbf{x}} = 1/(1 + \exp - 2\boldsymbol{\beta}'\tilde{x})$. The solution $\hat{\boldsymbol{\beta}}$ must satisfy (2.3-12). For the present case, $n_{10} = 35, n_{11} = 30, n_{12} = 34$, and using a Newton–Raphson procedure the solution is found to be $\hat{\boldsymbol{\beta}}' = (-1.36, .72, 1.45)$. The estimate of $\log[f_1(\mathbf{x})/f_2(\mathbf{x})]$ is then taken to be $2\hat{\boldsymbol{\beta}}'\tilde{x}$.

The minimum logit χ^2 estimates of $\boldsymbol{\beta}$ must minimize

$$\sum_{\mathbf{x}} \frac{n_1(\mathbf{x})n_2(\mathbf{x})}{n(\mathbf{x})}\left[\log\frac{n_1(\mathbf{x})}{n_2(\mathbf{x})} - 2\boldsymbol{\beta}'\tilde{x}\right]^2$$

under the assumption that

$$\log\frac{f_1(\mathbf{x})}{f_2(\mathbf{x})} = 2\boldsymbol{\beta}'\tilde{x}.$$

For the sample data given, the solution is $\hat{\boldsymbol{\beta}}' = (-0.93, 0.57, 1.10)$. Table 2.3-2 presents the estimated likelihood ratios and the correlation coefficient r of these estimates and the actual $f_1(\mathbf{x})/f_2(\mathbf{x})$ [weighted by $f(\mathbf{x})$].

Example 2.3-3   A Comparison of the Discriminant Function and Maximum-likelihood Approaches (Halperin, Blackwelder, and Verter [1971]).   In discussing a variety of approaches to estimating (2.3-10) we indicated that Halperin and colleagues reported results of a study comparing the estimates of the $\boldsymbol{\beta}$ coefficients by the discriminant-function and maximum-likelihood methods. To illustrate the differences and similarities of these

TABLE 2.3-3

GROUP FREQUENCIES FOR MALE SUBJECTS IN THE FRAMINGHAM
HEART STUDY

AGE (YR)	DEVELOPED CHD IN 12-YEAR PERIOD (Π_1)	NO CHD (Π_2)
29–39	40	749
40–49	88	742
50–62	130	656
All ages	258	2147

two approaches, we utilize some of the results reported in that study.

Halperin and colleagues used data originally presented in the Truett, Cornfield, and Kannel [1967] study in which a discriminant-function approach was taken in fitting 12-year incidence of coronary heart disease (CHD) data, collected in a Framingham, Massachusetts heart study, to a number of independent variables. The variables used for analysis are

X_1: Age (yr)

X_2: Serum cholesterol (mg/100 ml)

X_3: Systolic blood pressure (mm Hg)

X_4: Relative weight ($100 \times$ actual weight $\div$ median for sex–height group)

X_5: Hemoglobin (g/100 ml)

X_6: Cigarettes per day
(0 = never smoked
1 = less than 1 pack/day
2 = 1 pack/day
3 = more than 1 pack/day)

X_7: ECG (0 = normal; 1 = abnormal).

The Framingham data were analyzed for males aged 29–62, and the sample sizes for these groups were as shown in Table 2.3-3. However, although the authors considered a variety of situations with respect to the number and type of variables (dichotomous or continuous) involved in the analysis, we consider their results for only the variables as described in the preceding list.

Table 2.3-4 compares the discriminant-function estimates of parameters with the maximum likelihood estimates for the 12-year male CHD incidence data. In addition, the estimated standard errors of the respective estimates as well as the ratios of estimates to their standard errors are

TABLE 2.3-4

COMPARISON OF MAXIMUM-LIKELIHOOD (ML) AND DISCRIMINANT-FUNCTION (DF)
FITTING OF THE MULTIPLE LOGISTIC MODEL TO 12-YEAR CHD INCIDENCE: ESTIMATED
COEFFICIENTS AND THEIR STATISTICAL "SIGNIFICANCE"

	AGE (YR)		CHOLESTEROL (mg/100 ml)		SYSTOLIC BLOOD PRESSURE (mm Hg)		RELATIVE WEIGHT	
	ML	DF	ML	DF	ML	DF	ML	DF
Men 29–39								
Coefficient	0.1326	0.0920	0.0179	0.0231	0.0124	0.0219	0.0153	0.0139
Standard error	0.0738	0.0628	0.0038	0.0040	0.0106	0.0111	0.0132	0.0126
t	1.80	1.47	4.75	5.74	1.16	1.97	1.17	1.10
Men 40–49								
Coefficient	0.1216	0.1201	0.0070	0.0074	0.0068	0.0086	0.0257	0.0269
Standard error	0.0437	0.0413	0.0025	0.0027	0.0060	0.0063	0.0091	
t	2.78	2.91	2.77	2.77	1.13	1.38	2.81	3.01
Men 5(–62								
Coefficient	0.0711	0.0724	0.0087	0.0091	0.0143	0.0158	0.0071	0.0077
Standard error	0.0319	0.0307	0.0024	0.0024	0.0041	0.0043	0.0078	0.0076
t	2.23	2.36	3.58	3.82	3.45	3.67	0.91	1.01

	HEMOGLOBIN (dg/100 ml)		CIGARETTE SMOKING (CODED 0, 1,2,3)		ECG (CODED 0,1)		INTERCEPT	
	ML	DF	ML	DF	ML	DF	ML	DF
Men 29–39								
Coefficient	0.0025	0.0026	0.7315	0.5981	1.1208	1.2874	−16.9010	−17.6355
Standard error	0.0148	0.0136	0.1871	0.1436	0.6443	0.7994		
t	0.17	0.19	3.91	4.17	1.74	1.61		
Men 40–49								
Coefficient	−0.0010	−0.0011	0.4223	0.4336	0.7206	1.0525	−13.2573	−13.6995
Standard error	0.0098	0.0094	0.1031	0.0984	0.4009	0.4751		
t	−0.10	−0.12	4.09	4.41	1.80	2.22		
Men 50–62								
Coefficient	−0.0180	−0.0170	0.2743	0.2723	0.5695	0.7311	−7.9843	−8.6035
Standard error	0.0083	0.0078	0.0955	0.0922	0.2974	0.3369		
t	−2.17	−2.19	2.87	2.95	1.91	2.17		

Source. Halperin, Blackwelder, and Verter [1971].

shown. From the table it is apparent that in some instances the discriminant-function estimates of the β coefficients differ markedly from the maximum-likelihood estimates. The authors comment by suggesting that such differences might be expected for the ECG variable since it is dichotomous; however, they also point to the differences found for more continuous variables, such as systolic blood pressure in the 29-39-year age group. On the other hand, it is also apparent that the estimates by the two approaches are in substantial agreement in many cases and, in general, the ratios of coefficient to estimated standard error are more in accord than are the estimates.

By calculating $P\{\Pi_1(\text{CHD})|x\}$ for all male subjects in the sample and ordering the subjects on the basis of this estimated risk, it is possible to calculate the expected number of cases for each decile of risk and compare it with the observed number of cases. Note that in the maximum-likelihood approach the number of cases observed must necessarily equal the number of expected cases, whereas if the discriminant coefficients are used the number of observed and expected cases need not be identical.

TABLE 2.3-5

Comparison of Maximum-likelihood (ML) and Discriminant-function (DF) Fitting of the Multiple Logistic Model to 12-year CHD Incidence for Seven Risk Factors: Expected (Exp) and Observed (Obs) Numbers of Cases and Regression Coefficients of Observed on Expected

DECILES OF RISK	MEN 29–39 YR				MEN 40–49 YR				MEN 50–62 YR			
	ML		DF		ML		DF		ML		DF	
	EXP	OBS	EXP	OBS	EXP	OBS	EXP	OBS	EXP	OBS	EXP	OBS
10	18.66	19	21.94	18	24.10	27	26.11	26	30.11	32	31.82	32
9	7.80	8	7.66	8	15.21	14	15.38	15	20.70	19	21.29	17
8	4.67	3	4.06	4	11.45	6	11.38	6	16.86	19	16.88	23
7	2.93	5	2.63	4	8.95	11	8.79	11	13.78	11	13.57	8
6	2.02	3	1.82	4	7.41	9	7.13	9	11.72	10	11.43	10
5	1.46	1	1.28	1	6.18	5	5.91	4	10.29	11	9.83	14
4	1.07	1	0.92	1	5.12	8	4.84	9	8.64	10	8.29	7
3	0.74	0	0.64	0	4.22	3	3.99	3	7.50	8	7.08	10
2	0.44	0	0.39	0	3.27	4	3.03	4	6.07	4	5.71	3
1	0.21	0	0.18	0	2.09	1	1.89	1	4.33	6	4.04	6
Total	40.00	40	41.52	40	88.00	88	88.45	88	130.00	130	129.94	130
Regression coefficient	1.023		0.821		1.054		0.944		1.031		0.951	

Source. Halperin, Blackwelder, and Verter [1971].

Table 2.3-5 shows the comparison of observed and expected cases per decile for each method of estimation along with the regression coefficient of observed on expected by decile. It will be noticed that the observed for the two methods do not agree, and for the discriminant function approach the observed is not identical with the expected. The former result is a consequence of the fact that the two estimation procedures do not yield exactly the same ordering of the estimated individual risks. Note, also, that the maximum likelihood approach appears to give a better fit (as measured by the regression coefficient) for the 29–39-year age group, while for the 40–49 and 50–62-year age groups there seems to be little difference between the two methods.

To summarize, the final conclusion of this study was that if one is fitting the logistic model in order to isolate relevant risk factors, the discriminant-function approach works reasonably well, even in situations where the normality assumption is clearly violated. However, if one's main purpose is to estimate the actual magnitudes of parameters or of probabilities of events, then use of the discriminant-function method can in many instances yield misleading results.

2.3.3 Procedures Using Orthogonal Polynomials

The material presented in this section focuses on two approaches that utilize orthogonal functions to affect classification. In both approaches the multivariate binary density at the point $\mathbf{x}$ is expressed as a linear combination of orthogonal polynomials.

The Martin–Bradley Model. Martin and Bradley [1972] developed a class of probability models for multinomial distributions generated by p dichotomous random variables X_j, $j = 1,\ldots,p$. Probability models of the class (later referred to as *complete models*)

$$f_i(\mathbf{x}) = f(\mathbf{x})\big[1 + h(\mathbf{a}^{(i)}, \mathbf{x})\big] \qquad (2.3\text{-}13)$$

$i = 1, 2$, were considered where $h(\mathbf{a}^{(i)}, \mathbf{x})$ is a polynomial in the elements of $\mathbf{x}$ and the coefficients $\mathbf{a}^{(i)}$ are specific to Π_i. The function $f(\mathbf{x})$ is defined by

$$f(\mathbf{x}) = w_1 f_1(\mathbf{x}) + w_2 f_2(\mathbf{x}), \qquad w_1 + w_2 = 1, \qquad w_i \geqslant 0, \qquad i = 1,2 \quad (2.3\text{-}14)$$

where the weights w_i are regarded as arbitrary and assumed known if independent samples are available, or unknown but estimated if the sampling is from the mixed population.

The authors express $h(\mathbf{a}^{(i)}, \mathbf{x})$ in terms of orthogonal polynomials $\phi_\gamma(\mathbf{x})$

where

$$\phi_0(\mathbf{x}) = 1, \qquad \phi_j(\mathbf{x}) = 2x_j - 1, \qquad j = 1, 2, \ldots, p$$

$$\phi_\gamma(\mathbf{x}) = \prod_{j=1}^{k} \phi_{\gamma_j}(\mathbf{x}), \qquad \gamma = (\gamma_1, \gamma_2, \ldots, \gamma_k), \qquad \gamma_1 < \gamma_2, \ldots, \gamma_k$$

$$k = 2, 3, \ldots, p, \qquad \gamma_j \in \{1, 2, \ldots, k\}. \tag{2.3-15}$$

The complete set of 2^k values of γ is denoted by Γ_k, indicating all polynomial terms up to and including order k. The orthogonal property follows from

$$\sum_x \phi_\gamma(\mathbf{x}) \phi_\delta(\mathbf{x}) = 2^k \Delta(\gamma, \delta), \gamma, \delta \in \Gamma_k \tag{2.3-16}$$

where $\Delta(\gamma, \delta) = 1, 0$ as $\gamma =, \neq \delta$. Because the set of 2^k polynomials $\phi_\gamma(x)$, $\gamma \in \Gamma_k$, forms a basis for the set of all real-valued functions defined on the sample space generated by all the x_i values, it follows that for any set of probability functions $f_i(\mathbf{x})$, $i = 1, 2$, we may write

$$h(\mathbf{a}^{(i)}, \mathbf{x}) = \sum_{\gamma \in \Gamma_k} a_\gamma^{(i)} \phi_\gamma(\mathbf{x}). \tag{2.3-17}$$

The above representations show immediately that

$$h(\mathbf{a}^{(i)}, \mathbf{x}) = \frac{f_i(\mathbf{x}) - f(\mathbf{x})}{f(\mathbf{x})}$$

and

$$\mathbf{a}_\gamma^{(i)} = 2^{-k} \sum_x \phi_\gamma(\mathbf{x}) \frac{f_i(\mathbf{x}) - f(\mathbf{x})}{f(\mathbf{x})} \tag{2.3-18}$$

for $i = 1, 2$ and $\gamma \in \Gamma_k$, provided $f(\mathbf{x}) \neq 0$.

In the case of independent random samples available from Π_1 and Π_2, maximum-likelihood estimates for $f_i(\mathbf{x})$ are

$$\hat{f}_i(\mathbf{x}) = \frac{n_i(\mathbf{x})}{n_i} \tag{2.3-19}$$

$i = 1, 2$, where n_i is the sample size from Π_i and $n_i(\mathbf{x})$ is the frequency in state **x**. Further, with the weights w_1, w_2 assumed known

$$\hat{f}(\mathbf{x}) = w_1 \hat{f}_1(\mathbf{x}) + w_2 \hat{f}_2(\mathbf{x})$$

and

$$\hat{\mathbf{a}}_\gamma^{(i)} = 2^{-k} \sum_{\mathbf{x}} \phi_\gamma(\mathbf{x}) Y^{(i)}(\mathbf{x}) \tag{2.3-20}$$

where $Y^{(i)}(\mathbf{x}) = [\hat{f}_i(\mathbf{x}) - \hat{f}(\mathbf{x})]/\hat{f}(\mathbf{x})$, $i = 1$, 2, again provided that $\hat{f}(\mathbf{x}) \neq 0$. Similar representations hold when sampling from the mixed population. Problems clearly arise, however, with sparse data sets since $Y^{(i)}(\mathbf{x})$ is ill defined. On the other hand, if sparseness is not a problem, then a classification rule is defined in the usual way.

Indeed, when all 2^k parameters are estimated, then the induced classification rule is equivalent to the full multinomial rule. Hence the potentially useful models will depend on deletion of selected parameters in the complete expansion of $h(\mathbf{a}^{(i)}, \mathbf{x})$. Toward this end, there are essentially two ways of proceeding. First, Martin and Bradley propose fitting a reduced model of the form

$$f_i(\mathbf{x}) = f(\mathbf{x})\left[1 + h_s(\mathbf{a}^{(i)}, \mathbf{x})\right] \tag{2.3-21}$$

where s denotes a particular order of subset of polynomials, usually corresponding to main effects and low-order interactions. If we denote Γ_s as the set of polynomials of order no more than s, then

$$h_s(\mathbf{a}^{(i)}, \mathbf{x}) = \sum_{\gamma \in \Gamma_s} \mathbf{a}_\gamma^{(i)} \phi_\gamma(\mathbf{x}). \tag{2.3-22}$$

Various restrictions are imposed on $f(\mathbf{x})$ and $h_s(\mathbf{a}^{(i)}, \mathbf{x})$ so that all probabilities $f_i(\mathbf{x})$ are positive and sum to unity. The authors use maximum likelihood estimation; however, because the likelihood equations form a nonlinear system, closed estimates cannot be found, and hence iteration methods must be used. Of particular interest in terms of the classification problem is the potential usefulness of some low-order reduced models when data sets under analysis suffer from sparseness.

A second approach using the complete model follows the reasoning we applied in discussing rules resulting from a Bahadur representation of state probabilities. In particular, if we set all second- and higher-order interaction parameters to zero in $h(\mathbf{a}^{(i)}, \mathbf{x})$, then $f_i(\mathbf{x})$ may not be positive, nor will probabilities sum to unity. However, the likelihood-ratio rule might still be an effective classification procedure. For certain multinomial structures it has been found that reasonably good classification results when main effects are used in conjunction with first-order interaction parameters. Although this approach might yield reasonable error rates, like the full multinomial rule it will be of little value when many of the state frequencies are zero.

Before presenting an example, it should be noted that the polynomials defined by (2.3-15) are analogous to the independent variables of a 2^p-factorial in the analysis of variance. Therefore, in general, the $\mathbf{a}_\gamma^{(i)}$ may be interpreted in a similar fashion. For example, $a_0^{(i)}$ is a population characteristic analogous to a group mean, $a_j^{(i)}$ corresponds to the main effect for the jth factor, $a_{j\cdot k}^{(i)}, j \neq k$, corresponds to the interaction between the jth and the kth factors, and so on. Translating this to the classification problem, $a_j^{(i)}$ measures the ability of the jth dichotomous variate as an indicator to the ith population, $a_{j,k}^{(i)}$ measures the joint ability of the jth and kth variates as an indicator to the ith population, and so on. Hence it is possible to describe a situation where a variate is found not contributing to classification when considered alone, that is, $a_j^{(i)} = 0$; however, when considered jointly with another variate their contribution is found to be significant.

Example 2.3-4   Data on Detergent Preferences. The data of Table 2.3-6 are a modified version of the data originally collected by Reis and Smith [1963]. They result from an experiment in which 1008 people were given two brands of detergent, X and M, and subsequently asked questions regarding four points, corresponding to the four variables: (a) water softness—soft, medium, or hard, (b) previous use of brand M—yes or no, (c) water temperature—high or low, and (d) brand preference—X or M. The frequency distributions given in Table 2.3-6 reflect two modifications, namely, that we have used variable (d) as the grouping variable and that

TABLE 2.3-6

Observed Frequency Distributions for Data
on Detergent Preference

STATE $(x_1 x_2 x_3)$	BRAND PREFERENCE			
	$X(\Pi_1)$		$M(\Pi_2)$	
	OBSERVED	RELATIVE	OBSERVED	RELATIVE
1 1 1	19	0.056	29	0.089
1 1 0	57	0.168	49	0.151
1 0 1	29	0.086	27	0.083
1 0 0	63	0.186	53	0.163
0 1 1	24	0.071	43	0.132
0 1 0	37	0.109	52	0.160
0 0 1	46	0.124	30	0.093
0 0 0	68	0.200	42	0.129
	339	1.000	325	1.000

we treat variable (a) as a binary response, corresponding to soft or hard. Hence, we now have three binary variables generating $2^3 = 8$ possible states, and respective samples of size 339 and 325.

Note that in addition to the relative frequencies, Table 2.3-6 gives the estimate of $\hat{f}(\mathbf{x})$, where the weights w_i, $i = 1$, 2 are assumed equal. In evaluating the Martin and Bradley model we take the second approach to using the complete model in that we report results for the case where no more than main effect terms are used and when terms no more than first-order interactions are assumed.

Using the Martin and Bradley model the $\mathbf{a}_\gamma^{(i)}$ parameters evaluate to

Preference	$\hat{a}_0$	$\hat{a}_1$	$\hat{a}_2$	$\hat{a}_3$	$\hat{a}_{12}$	$\hat{a}_{13}$	$\hat{a}_{23}$	$\hat{a}_{123}$
X	-0.0279	0.0044	-0.1385	-0.0647	0.0751	-0.0187	-0.0345	-0.0236
M	0.0279	-0.0044	0.1385	0.0647	-0.0751	0.0187	0.0345	0.0236

Thus we obtain the following misclassification probabilities: (a) 0.4232 for the complete model, (b) 0.4301 for the main effects model, and (c) 0.4259 for the first-order interaction model.

Note first that since we assumed $w_1 = w_2 = 0.5$, $\hat{a}_\gamma^{(1)} = -\hat{a}_\gamma^{(2)}$; however, the sign is, in general, arbitrary. In terms of the magnitude of the $\hat{a}_\gamma^{(i)}$ coefficients, we see that $\hat{a}_2^{(i)}$ is relatively large and, therefore, it seems that users tend to prefer the brand they know (this finding is consistent with other analyses of the data; see Bishop, Fienberg, and Holland [1975]). Finally, use of either the main effects or first-order interaction model yields results only slightly different from the complete model.

The Kronmal–Ott–Tarter Model.   As before, suppose $\mathbf{X} = (X_1, X_2, \ldots, X_p)$ is a multivariate binary vector with an associated sample space consisting of 2^p points. Let these points be numbered by a binary index $\mathbf{r}$ and consider the orthogonal functions

$$\psi_\mathbf{r}(\mathbf{x}) = (-1)^{\mathbf{x'r}} \qquad (2.3\text{-}23)$$

where $\mathbf{x'r} = \sum_j x_j r_j$. The orthogonality of these functions follows by observing that

$$\sum_\mathbf{x} \psi_\mathbf{r}(\mathbf{x})\psi_\mathbf{j}(\mathbf{x}) = \begin{array}{l} \sum_x (-1)^{2\mathbf{x'r}} = 2^p \quad \text{for} \quad \mathbf{r} = \mathbf{j} \\ \\ \sum_x (-1)^{\mathbf{x'(r+j)}} = 0 \quad \text{for} \quad \mathbf{r} \neq \mathbf{j}. \end{array} \qquad (2.3\text{-}24)$$

Kronmal and Tarter [1968] represent the probability associated with a point x by

$$f(\mathbf{x}) = 2^{-p} \sum_{\mathbf{r}} d_{\mathbf{r}} \psi_{\mathbf{r}}(\mathbf{x}) \tag{2.3-25}$$

where d_j is a parameter associated or determined by noting that

$$\psi_{\mathbf{j}}(\mathbf{x}) f(\mathbf{x}) = 2^{-p} \sum_{\mathbf{r}} d_{\mathbf{r}} \psi_{\mathbf{r}}(\mathbf{x}) \psi_{\mathbf{j}}(\mathbf{x}) \tag{2.3-26}$$

and hence

$$\sum_{\mathbf{x}} \psi_{\mathbf{j}}(\mathbf{x}) f(\mathbf{x}) = 2^{-p} \sum_{\mathbf{r}} d_{\mathbf{r}} \sum_{\mathbf{x}} \psi_{\mathbf{r}}(\mathbf{x}) \psi_{\mathbf{j}}(\mathbf{x})$$

$$= d_{\mathbf{j}} = E\big[\psi_{\mathbf{j}}(\mathbf{x})\big]. \tag{2.3-27}$$

Based on a random sample of size n, let $n(\mathbf{x})$ be the frequency of the state defined by x. Then the maximum likelihood estimate of $f(\mathbf{x})$ can be written as

$$\hat{f}(\mathbf{x}) = 2^{-p} \sum_{\mathbf{r}} \hat{d}_{\mathbf{r}} \psi_{\mathbf{r}}(\mathbf{x}) \tag{2.3-28}$$

where $\hat{d}_{\mathbf{r}} = \sum_{\mathbf{x}} \psi_{\mathbf{j}}(\mathbf{x}) n(\mathbf{x}) / n$.

As expressed, $f(\mathbf{x})$ is a linear function of $2^p - 1$ independent parameters interpretable in terms of individual contribution of the binary variables, or joint contribution of first order or higher; similar, as was the case in the Martin–Bradley model, to that of the familiar analysis of variance models. When all the $\hat{d}_{\mathbf{r}}$ estimates are included in the representation of $\hat{f}(\mathbf{x})$ in (2.3-28), then $\hat{f}(\mathbf{x}) = n(\mathbf{x})/n$. To achieve parsimony, that is reduce the number of parameters to be estimated and yet have the estimates of state probabilities and true probabilities close in some reasonable sense, Kronmal and Tarter have shown using mean summed squared error

$$E \sum_{\mathbf{x}} \big(f(\mathbf{x}) - \hat{f}(\mathbf{x})\big)^2 \tag{2.3-29}$$

as a criterion of fit that the increase in error due to inclusion of the rth term, namely, d_r, in the representation of $f(\mathbf{x})$ is given by

$$\frac{2^{-p}\big[1 - (n+1)d_{\mathbf{r}}^2\big]}{n} \tag{2.3-30}$$

and is estimated by

$$\frac{2^{-p}\left[2-(n+1)\hat{d}_r^2\right]}{n-1} \tag{2.3-31}$$

It follows, therefore, that inclusion of $\hat{d}_r$ in $\hat{f}(\mathbf{x})$ leads to a decrease in error if (2.3-31) is negative; that is, if

$$\hat{d}_r^2 > \frac{2}{n+1}. \tag{2.3-32}$$

Returning now to the development of a classification rule that follows Goldstein [1977], suppose independent samples of size n and m are available from Π_1 and Π_2, respectively. Suppose further that an observation with pattern $\mathbf{x}$ is to be classified. Assuming equal prior probabilities and denoting by f_1 and f_2 the underlying multinomial densities respectively associated with Π_1 and Π_2, it follows that the optimal classification rule using the representation discussed above is to classify $\mathbf{x}$ into $\Pi_1(\Pi_2)$ if

$$\sum_r \psi_r(\mathbf{x})d_{1,r} > (<) \sum_r \psi_r(\mathbf{x})d_{2,r} \tag{2.3-33}$$

and randomly assign otherwise, where the parameter sets $\{d_{i,r}\}$ are those associated with Π_1 and Π_2, respectively.

If all parameters are estimated, the sample-based rule is simply the rule given in (2.3-33) with the parameter sets $\{d_{i,r}\}$ replaced by their estimates $\{\hat{d}_{i,r}\}$, $i=1, 2$. However, as the discussion above indicates, it may not be necessary to estimate all the parameters. To illustrate this, suppose since our sample-based rule is determined by the sign of $\hat{f}_1(\mathbf{x}) - \hat{f}_2(\mathbf{x})$ (i.e., a positive difference requires that $\mathbf{x}$ be allocated Π_1, and a negative difference assigns $\mathbf{x}$ to Π_2), whereas the optimal rule is based on the difference $f_1(\mathbf{x}) - f_2(\mathbf{x})$, it seems that we should only incorporate estimates of parameters in defining the rule that minimizes the expected summed difference between the two. Mathematically, this translates into minimizing

$$E\sum_{\mathbf{x}}\left[\hat{f}_1(\mathbf{x}) - \hat{f}_2(\mathbf{x}) - (f_1(\mathbf{x}) - f_2(\mathbf{x}))\right]^2. \tag{2.3-34}$$

However, since $\hat{f}_i(\mathbf{x})$ is an unbiased estimate of $f_i(\mathbf{x})$, $i=1, 2$, and $\hat{f}_1(\mathbf{x})$ and $\hat{f}_2(\mathbf{x})$ are independent (since independent samples are available), it follows that (2.3-34) is minimized if

$$E\sum_{\mathbf{x}}\left(\hat{f}_i(\mathbf{x}) - f_i(\mathbf{x})\right)^2$$

$i = 1, 2$ are individually minimized. Therefore, the inclusion of $\hat{d}_{1,r}$ and/or $\hat{d}_{2,r}$ into the sample-based estimate of the rule given in (2.3-33) leads to a decrease in the error between the optimal rule and its estimate if

$$\hat{d}_{1,r}^2 > \frac{2}{n+1}$$

and/or

$$\hat{d}_{2,r}^2 > \frac{2}{m+1} \tag{2.3-35}$$

Now let

$$S_{1,r} = \left\{ \mathbf{r} \,\middle|\, \hat{d}_{1,r}^2 > \frac{2}{n+1} \right\}$$

$$S_{2,r} = \left\{ \mathbf{r} \,\middle|\, \hat{d}_{2,r}^2 > \frac{2}{m+1} \right\} \tag{2.3-36}$$

and define the indicator random variables

$$I(S_{i,r}) = \begin{matrix} 1 & \text{if} & S_{i,r} \quad \text{occurs} \quad (i=1,2); \\ 0 & \text{if not.} \end{matrix} \tag{2.3-37}$$

Accordingly, we are led to the following sample-based rule; namely, to classify $\mathbf{x}$ into $\Pi_1(\Pi_2)$ if

$$\sum_{\mathbf{r}} \psi_{\mathbf{r}}(\mathbf{x}) \hat{d}_{1,r} I(S_{1,r}) > (<) \sum_{\mathbf{r}} \psi_{\mathbf{r}}(\mathbf{x}) \hat{d}_{2,r} I(S_{2,r}), \tag{2.3-38}$$

and randomly assign if otherwise.

Note that the rule given in (2.3-38) is equivalent to the full multinomial rule for independent samples if $I(S_{i,r}) = 1$ for $i = 1, 2$ and all $\mathbf{r}$.

Ott and Kronmal [1976] utilized the density estimation methods of Kronmal and Tarter [1968] in developing classification methods assuming sampling from the mixed population. Their basic classification rule (which they refer to as the *basic Fourier procedure*) and their *unbiased difference estimate method* are very similar to the rule given in (2.3-38). In addition, they consider two other methods for defining procedures based on the basic density estimation representation and compare all methods through Monte Carlo sampling experiments. As a general statement, no one procedure appears to be superior on the basis of the population structures that were employed to generate the data.

A limited but informative discussion of the relationship between the orthogonal functions presented above and discrete Fourier transforms is given in the appendix of Ott and Kronmal's paper. In particular, their introduction of the use of the fast Hadamard transform (Shum and Elliott [1972]) is most instructive and helpful since without these ideas practical implementation of the rule given in (2.3-38) would be computationally difficult for large multinomials. The interested reader is referred to the bibliography of the Ott and Kronmal paper for details on implementation.

Example 2.3-5 Data on the Behavioral Consequences Following Hypoxic Trauma. As an example of the calculation of the estimates $\{d_{i,r}\}$, $i=1, 2$ and the subsequent rule, we consider a subset of data originally reported in Martin and Bradley [1972] dealing with the behavioral consequences following hypoxic trauma (damage to an infant during and shortly after birth caused by oxygen deficiency). The data, given in Table 2.3-7, were supplied by Joan C. Martin and Celia Lamper of the Duke University Medical Center. The two populations in question are defined by infants with Apgar scores of 7 or below (Π_1) and those with normal Apgar scores (Π_2). The score is an index of the level of physiological functioning based on symptoms of the infant observed immediately following birth. For purposes of illustration, we consider only three dichotomous variables: (a) race, (b) suggestive or nonsuggestive previous medical history of mother, and (c) infant's first breath before or after 5 sec. We further assume independent samples of size 106 from Π_1 and 113 from Π_2.

For convenience and simplicity of notation, we denote the parameter for $\Pi_i, i=1,2$, associated with state (000) as $d_{j,1}$ (001) as $d_{j,2},\ldots,$ and (111) as

TABLE 2.3-7

OBSERVED FREQUENCY DISTRIBUTIONS FOR NORMAL INFANTS AND THOSE HAVING SYMPTOMS SUGGESTIVE OF DAMAGE

$(x_1x_2x_3)$	SUGGESTIVE OF DAMAGE (Π_1)		NORMAL (Π_2)	
	OBSERVED	RELATIVE	OBSERVED	RELATIVE
0 0 0	24	0.226	31	0.274
0 0 1	0	0.000	0	0.000
0 1 0	48	0.453	36	0.319
0 1 1	3	0.028	0	0.000
1 0 0	8	0.075	22	0.195
1 0 1	0	0.000	0	0.000
1 1 0	21	0.198	24	0.212
1 1 1	2	0.019	0	0.000
	106	1.000	113	1.000

$d_{j,8}$. It follows that

$$\hat{d}_{1,1} = 0.226 \;\; +0.000 \;\; +0.453 \;\; +0.028 \;\; +0.075 \;\; +0.000 \;\; +0.198 \;\;\;\; 0.019 \;\; = \;\;\;\; 0.999$$
$$\hat{d}_{1,2} = 0.226 \;\; -0.000 \;\; +0.453 \;\; -0.028 \;\; +0.075 \;\; -0.000 \;\; +0.198 \;\; -0.019 \;\; = \;\;\;\; 0.905$$
$$\hat{d}_{1,3} = 0.226 \;\; +0.000 \;\; -0.453 \;\; -0.028 \;\; +0.075 \;\; +0.000 \;\; -0.198 \;\; -0.019 \;\; = -0.397$$
$$\hat{d}_{1,4} = 0.226 \;\; -0.000 \;\; -0.453 \;\; +0.028 \;\; +0.075 \;\; -0.000 \;\; -0.198 \;\; +0.019 \;\; = -0.303$$
$$\hat{d}_{1,5} = 0.226 \;\; +0.000 \;\; +0.453 \;\; +0.028 \;\; -0.075 \;\; -0.000 \;\; -0.198 \;\; -0.019 \;\; = \;\;\;\; 0.415$$
$$\hat{d}_{1,6} = 0.226 \;\; -0.000 \;\; +0.453 \;\; -0.028 \;\; -0.075 \;\; +0.000 \;\; -0.198 \;\; +0.019 \;\; = \;\;\;\; 0.397$$
$$\hat{d}_{1,7} = 0.226 \;\; +0.000 \;\; -0.453 \;\; -0.028 \;\; -0.075 \;\; -0.000 \;\; +0.198 \;\; +0.019 \;\; = -0.113$$
$$\hat{d}_{1,8} = 0.226 \;\; -0.000 \;\; -0.453 \;\; +0.028 \;\; -0.075 \;\; +0.000 \;\; +0.198 \;\; -0.019 \;\; = -0.095$$

Similar calculations show:

$$\hat{d}_{2,1} = \;\;\; 1.000$$
$$\hat{d}_{2,2} = \;\;\; 1.000$$
$$\hat{d}_{2,3} = -0.062$$
$$\hat{d}_{2,4} = -0.062$$
$$\hat{d}_{2,5} = \;\;\; 0.186$$
$$\hat{d}_{2,6} = \;\;\; 0.186$$
$$\hat{d}_{2,7} = -0.028$$
$$\hat{d}_{2,8} = -0.028.$$

According to criterion to (2.3-24) the estimates $\hat{d}_{1,7}$, $\hat{d}_{1,8}$, $\hat{d}_{2,3}$, $\hat{d}_{2,4}$, $\hat{d}_{2,7}$, and $\hat{d}_{2,8}$ can be set equal to zero.

Following the discussion above, direct calculations show that

	(000)	(001)	(010)	(011)	(100)	(101)	(110)	(111)
$\hat{f}_1$	0.252	0.002	0.427	0.026	0.049	0.000	0.224	0.021
$\hat{f}_2$	0.297	0.000	0.026	0.000	0.204	0.000	0.204	0.000

One of the interesting observations to be made from these calculations is that under Π_1 the state (001), which had a frequency estimate (maximum likelihood) of 0.000 (because of no available data for that state), results now in a nonzero estimate and hence enables us to effect a classification. Therefore, we see that deleting parameter estimates in defining the density estimate $\hat{f}(\mathbf{x})$ can actually effectuate more reasonable classification rules. Finally, note that the rule given in (2.3-38) correctly classifies 0.590 of the observations and is slightly worse than the full multinomial rule, which correctly classifies 0.600 of the observations.

2.4   A PROCEDURE BASED ON A DISTRIBUTIONAL DISTANCE

In the previous section of this chapter, classification rules were for the most part formed through approximations to and estimates of multinomial densities. The procedures to be discussed in this section are derived through a different approach in that likelihood ratios are not used as a basis for classification.

Toward this end, suppose $F_1 = \{p_j\}$ and $F_2 = \{q_j\}$, $j = 1,2,\ldots,s$, are two discrete distributions defined on the same space. In a series of papers Matusita [1954, 1955, 1957] used a measure of distance between F_1 and F_2 given by

$$\|F_1 - F_2\|^2 = \sum_{j=1}^{s} \left(\sqrt{p_j} - \sqrt{q_j} \right)^2 \tag{2.4-1}$$

to derive classification rules. His approach also implicitly relies on a measure of affinity defined by

$$\rho(F_1, F_2) = \sum_{j=1}^{s} \sqrt{p_j q_j} \tag{2.4-2}$$

which is related to $\| \ \|$ by

$$\|F_1 - F_2\|^2 = 2(1 - \rho). \tag{2.4-3}$$

If n and m independent observations are taken from F_1 and F_2, respectively, let $S_n = \{n_i/n\}$ and $S_m = \{m_i/m\}$ be the derived empirical distributions. Two fundamental properties that form the basis for most of the large sample results relating to performance of his classification rules are:

if $F_1 = F_2$, then

$$P\{\|S_n - S_m\| < \theta\} = P\left\{\rho(S_n, S_m) > 1 - \frac{\theta^2}{2}\right\}$$

$$\geq 1 - \frac{s}{\theta^2}\left(\frac{1}{\sqrt{n}} + \frac{1}{\sqrt{m}}\right)^2$$

$$\text{or} \doteq P\{\chi^2_{(s-1)} < n\theta^2\} P\{\chi^2_{(s-1)} < m\theta^2\} \tag{2.4-4}$$

where $\chi^2_{(t)}$ represents a χ^2 random variable with t degrees of freedom:

if $\|F_1 - F_2\| \geqslant \delta_0$, then

$$P\{\|S_n - S_m\| > \theta\} = P\left\{\rho(S_n, S_m) < 1 - \frac{\theta^2}{2}\right\}$$

$$\geqslant 1 - \frac{s-1}{(\delta_0 - \theta)^2}\left(\frac{1}{\sqrt{n}} + \frac{1}{\sqrt{m}}\right)^2$$

or $\doteq P\left\{\chi^2_{(s-1)} < n(\delta_0 - \theta)^2\right\}P\left\{\chi^2_{(s-1)} < m(\delta_0 - \theta)^2\right\}.$ (2.4-5)

Suppose now that l observations are to be classified in total into F_1 or F_2. Further, let $S_l = \{l_j/l\}$ be the derived empirical distribution. Matusita [1955] considers the following classification rule: Classify all observations into F_1 if

$$\|S_n - S_l\| \leqslant \|S_m - S_l\|$$ (2.4-6)

and into F_2 otherwise. In the same paper in which (2.4-6) is discussed the following consistency property is established: when all l observations belong to F_1

$$P\{\|S_n - S_l\| \leqslant \|S_m - S_l\|\} \geqslant \left(1 - \frac{16(s-1)}{nd^2}\right)\left(1 - \frac{16(s-1)}{md^2}\right)\left(1 - \frac{16(s-1)}{ld^2}\right)$$

(2.4-7)

for n, m, and l greater than $16(s-1)/d^2$ and

$$P\{\|S_n - S_l\| \leqslant \|S_m - S_l\|\}$$

$$\geqslant \left[1 - \frac{16(s^2+s-1)}{(nd^2)^2}\right]\left[1 - \frac{16(s^2+s-1)}{(md^2)^2}\right]\left[1 - \frac{16(s^2+s-1)}{(ld^2)^2}\right],$$ (2.4-8)

for n, m, l greater than $4(s^2+s-1)^{1/2}/d^2$ and k, and

$$P\{\|S_n - S_l\| \leqslant \|S_m - S_l\|\}$$

$$\doteq P\left\{\chi^2_{(s-1)} < 4n\frac{d^2}{16}\right\}P\left\{\chi^2_{(s-1)} < 4m\frac{d^2}{16}\right\}P\left\{\chi^2_{(s-1)} < 4l\frac{d^2}{16}\right\}$$ (2.4-9)

for sufficiently large n, m, l where $\chi^2(s-1)$ denotes a random variable having a χ^2 distribution with $s-1$ degrees of freedom.

A straightforward analysis of the rule given in (2.4-6) immediately shows that when $l=1$, the usual case in the classification problem, it and the full multinomial rule are equivalent. Hence Matusita's rule suffers from the same difficulties that plague the full multinomial rule. However, it does provide an approach to the problem of simultaneously classifying more than one observation, but it requires that they all be assigned to one population. Further, (2.4-7–9) do establish that the probability of erring tends to zero for rule (2.4-6) provided that the sample sizes increase without bound.

In a more recent paper Dillon and Goldstein [1978] proposed using the distributional distance (2.4-1) in a different manner for the purpose of defining a classification rule. Under the same setup as above but with one observation to be classified, the following rule was offered:

Classify into F_1 if $\|S_{n+1} - S_m\| > \|S_n - S_{m+1}\|$

Classify into F_2 if $\|S_{n+1} - S_m\| < \|S_n - S_{m+1}\|$

Randomly allocate if $\|S_{n+1} - S_m\| = \|S_n - S_{m+1}\|$ (2.4-10)

where by S_{n+1} we mean the empirical distribution based on a sample of size $n+1$, similarly for S_{m+1}. In words, the rule simply states that if assigning the observation to F_1 results in greater sample-based distributional distance than if the observation had been placed in F_2, then there should be classification into F_1.

Rewriting the first inequality in (2.4-10) yields the rule of classifying the observation into F_1 if

$$\sum_{j=1}^{s} \left(\sqrt{n_j^*/(n+1)} - \sqrt{m_j/m} \right)^2 > \sum_{j=1}^{s} \left(\sqrt{n_j/n} - \sqrt{m_j^*/(m+1)} \right)^2 \quad (2.4\text{-}11)$$

where if the observation to be classsified is a member of state k, then $n_j^*(m_j^*) = n_j(m_j)$ for $j \neq k$ and $(n_j^*)(m_j^*) = n_j + 1(m_j + 1)$ for $j = k$. If the observation is a member of state k, then the inequality in (2.4-11) reduces to

$$\sqrt{(n_k+1)/(n+1)} \, \sqrt{m_k/m} + \sum_{j \neq k} \sqrt{n_j/(n+1)} \, \sqrt{m_j/m}$$

$$< \sqrt{n_k/n} \, \sqrt{(m_k+1)/(m+1)} + \sum_{j \neq k} \sqrt{n_j/n} \, \sqrt{m_j/(m+1)} . \quad (2.4\text{-}12)$$

It now follows that with equal sample sizes $n = m$, (2.4-12) again reduces to the full multinomial rule. However, for unequal sample sizes the rules are distinct. In general, algebraic manipulation of (2.4-10) results in the following equivalent inequality:

$$\frac{[m_k(n_k+1)]^{1/2} + \sum_{j \neq k} [n_j m_j]^{1/2}}{[n_k(m_k+1)]^{1/2} + \sum_{j \neq k} [n_j m_j]^{1/2}} < \left[\frac{m(n+1)}{n(m+1)} \right]^{1/2} \qquad (2.4\text{-}13)$$

Note that the ratio $m(n+1)/n(m+1)$ is >1 if $n < m$, $=1$ if $n = m$, and <1 if $n > m$.

As we discussed earlier, one of the troublesome problems that frequently arises in multinomial classification involves the handling of sparse states. In particular, one difficulty in dealing with sparseness involves the probable nonequivalence of zeros; that is, a zero in state j from F_1 may really represent a greater theoretical frequency than a nonzero in state j from F_2, especially if n is disproportionately small with respect to m. The usual full multinomial rule stipulating that if the observation belongs to state j, then allocate it to F_1 if $n_j/n > m_j/m$ is insensitive to this problem since, all other things remaining equal, as soon as state j is zero in F_1 and nonzero in F_2, then X is allocated to F_2.

On the other hand, the rule as expressed in (2.4-13) is more sensitive to this issue. To see this, suppose $n < m$, so that $m(n+1)/n(m+1) > 1$, and $n_k = 0$ but $m_k \neq 0$. Then the rule allows us to classify the observation in question to F_1 if

$$\sqrt{m_k} < \sum_{j \neq k} \sqrt{n_j m_j} \left\{ \left[\frac{m(n+1)}{n(m+1)} \right]^{1/2} - 1 \right\}. \qquad (2.4\text{-}14)$$

Example 2.4-1 A Numerical Illustration. We consider a hypothetical eight-state problem, the data of which are shown in Table 2.4-1, to illustrate our point. Notice first that the sample size from Π_1 is disproportionately small relative to the sample size from Π_2. Also, for states (111), (100), and (001) no data are available in Π_1, and hence any future observations from these states would automatically be assigned to Π_2 with use of the usual full multinomial rule.

However, contrary to the full multinomial rule, we have posited that use of the inequality given in (2.4-14) can potentially lead to the assignment of future observations with $n_k = 0 < m_k$ into Π_1. Indeed, straightforward

TABLE 2.4-1

OBSERVED FREQUENCY DISTRIBUTIONS (HYPOTHETICAL)
FOR DATA ILLUSTRATING RULE (2.4-14)

STATE $(x_1x_2x_3)$	Π_1 OBSERVED	Π_1 RELATIVE	Π_2 OBSERVED	Π_2 RELATIVE
1 1 1	0	0.000	1	0.004
1 1 0	1	0.040	20	0.073
1 0 1	3	0.120	100	0.364
1 0 0	0	0.000	1	0.004
0 1 1	5	0.200	80	0.290
0 1 0	12	0.480	12	0.043
0 0 1	0	0.000	1	0.004
0 0 0	4	0.160	60	0.218
	25	1.000	275	1.000

calculations using inequality (2.4-14), which are shown in Table 2.4-2, verify this result. In general, the inequality given in (2.4-14) will be satisfied provided that the samples are disproportionate in size and m_k is small relative to the other m_j values. Therefore, we see that the distance rule assists in an area where the full multinomial rule has no value.

Another problem arising with the analysis of samples of grossly unequal size concerns the performance obtained in the smaller population group. That is, although in situations where the sample from one population is

TABLE 2.4-2

SUMMARY OF CALCULATIONS FOR RULE (2.4-14)

STATE $(x_1x_2x_3)$	Π_1 n_j	Π_2 m_j	$\dfrac{\sqrt{m_j(n_j+1)} + c'}{\sqrt{n_j(m_j+1)} + c}$	$[m(n+1)/n(m+1)]^{1/2}$	ALLOCATION
1 1 0	0	1	1.014	1.018	Π_1
1 1 0	1	20	1.025	1.018	Π_2
1 0 1	3	100	1.053	1.018	Π_2
1 0 0	0	1	1.014	1.018	Π_1
0 1 1	5	80	1.025	1.018	Π_2
0 1 0	12	12	1.000	1.018	Π_1
0 0 1	0	1	1.014	1.018	Π_1
0 0 0	4	60	1.024	1.018	Π_2

$t_c = \Sigma_{j\neq k} \sqrt{n_j m_j}$.

very much larger than that of the other the total misclassification probability is likely to be good, almost all of the observations from the smaller population will be misclassified into the larger population. This is particularly troublesome since it is reasonable to expect that in many applications the practitioner may often be more interested in the smaller group. The last example of this chapter illustrates the relative efficacy of the distance procedure in analyzing samples of disproportionate size.

Example 2.4-2 Shopper Characteristics and Selected Modes of Shopping Data. Data collected in a 1976 unpublished study designed to investigate the relationship between shopper characteristics, and selected modes of shopping for a durable equipment item are used to illustrate the relative efficacy of the distance method over the full multinomial rule in situations of disproportionate sample sizes. A national sample of 5900 power-tool owners, selected randomly from warranty cards and manufacturer drop shipment records of recent purchasers, were sent a mail questionnaire during spring 1976. A total of 984 usable questionnaires were returned. Included in the data collected were the actual shopping outlets selected. For the purposes of this illustration, we consider only a subset of the available sample, namely, those individuals selecting an in-home shopping mode—a catalog—and those shopping from a specific in-store alternative—a discount store. Of the 254 respondents who selected one of these two distinct types of shopping modes, 54 indicated that they shopped at a discount store, and the remaining 200 indicated that their purchase was initiated via catalog.

Eight variables were selected to form the basis for classification. A description of these variables can be found in Table 2.4-3. Note that of the eight predictors, six variables (home ownership, type of residence, marital status, education, use in business, and number of power tools owned) are dichotomous, whereas the remaining two variables (number of children and family income) are trichotomous. The observed state frequency distributions are not reported since with six dichotomous and two trichotomous predictors there is a total of 576 possible response patterns.

Table 2.4-4 presents summary results, in the form of a classification matrix, for the full multinomial, Fisher's LDF, and distance procedures (we report results for the Fisher LDF simply for illustrative purposes). Comparison of the results requires that measures of performance can be developed. Until now we have simply reported a procedure's total efficiency; that is, the resultant total misclassification probability. However, in evaluating several classification rules it seems useful to examine whether the performance of each method surpasses some standard of comparison. For example, we might consider the performance of a particular classification rule vis-à-vis assigning all observations to the largest

TABLE 2.4-3
VARIABLE DESCRIPTIONS

VARIABLE	DESCRIPTION
Home ownership (X_1)	Own
	Rent
Type of residence (X_2)	Single-family unit
	Multiple-family unit
Marital status (X_3)	Married
	Single, widowed, or divorced
Number of children (X_4)	Less than two children
	Two to four children
	Five children or more
Education (X_5)	No college
	Some college or above
Family income (X_6)	Less than $10,000 a year
	$10,000–19,999 a year
	$20,000 a year or more
Use in business (X_7)	Yes, I use
	No, I don't use
Number of tools owned (X_8)	Less than five tools
	Five tools or more

TABLE 2.4-4
CLASSIFICATION RESULTS FOR THE FULL, LDF, AND DISTANCE PROCEDURES

		PREDICTED						
		FULL		LDF		DISTANCE		
		1	2	1	2	1	2	TOTAL
actual	1	19	35	4	50	43	11	54
	2	4	196	2	198	45	155	200
		23	231	6	248	88	166	

48

population. Denoting c_{max} as the maximum chance model and letting α be the proportion of individuals belonging to Π_1 and $1 - \alpha$ the proportion of individuals in Π_2, then

$$c_{max} = \max{(\alpha, 1 - \alpha)}.$$

On the other hand, we might compare the performance of a given classification rule to what would have resulted if observations were randomly assigned to the populations with probabilities equal to the sample group frequencies. Hence, denoting c_{prop} as the proportional chance model, we have

$$c_{prop} = \alpha^2(1 - \alpha)^2.$$

For the interested reader, a more detailed discussion of these two rather heuristic chance classification schemes can be found in Morrison [1969].

Returning now to Table 2.4-4 we see that there appears to be some difficulty in interpreting the classification results. According to the table, the LDF misclassifies only 52 observations, but 50 of these come from Π_1, the smaller of the two populations. Therefore, whereas its total efficiency ($202/254 = 0.795$) is satisfactory as compared to either the distance method ($198/254 = 0.780$), the maximum chance model ($c_{max} = 0.787$), or the proportional chance model ($c_{prop} = 0.665$), it is of little value in identifying discount shoppers. Similarly, use of the full multinomial procedure yields better total performance ($215/254 = 0.846$) than the other methods and compares favorably to the maximum and chance models, but compared to the distance method it does relatively poorly in the smaller population group.

Clearly, what we are arguing is that in situations of disproportionate sample sizes total efficiency may not be the best estimate of discriminatory efficiency. Furthermore, in cases such as this, analysis of the individual group classifications is warranted, with particular emphasis on the accuracy of classification obtained in the smaller group. This translates into examining the conditional efficiency. For example, it seems most important to ask what the probability is of correctly identifying an observation given its group membership. That is, given that an individual is a discount (catalog) shopper, what is the probability of classifying the individual correctly? These probabilities are estimated from the classification matrix by the proportion of discount shoppers correctly classified out of the total number of discount shoppers and the proportion of catalog shoppers correctly classified out of the total number of catalog shoppers, respectively. In our case the values are: (a) 0.352 (19/54) and 0.980

(196/200) for the full multinomial; (b) 0.074 (4/54) and 0.990 (198/200) for the LDF; and (c) 0.796 (43/54) and 0.775 (155/200) for the distance method. Using the proportional chance model, the conditional chance probabilities are 0.213 (54/200) and 0.787 (200/254). Hence we see that the distance method correctly classifies 79.6% of the discount shoppers, which is far superior to the full multinomial and LDF procedures and, in addition, its conditional efficiency in the catalog-shopper group is only slightly different from the conditional proportional chance model.

CHAPTER 3

Error Rates and the Problem of Bias

3.1 INTRODUCTION

Our purpose in this chapter is to give a somewhat detailed discussion of the relationship between the optimum Bayes error and certain misclassification errors that evolve through the use of the sample-based full multinomial rule. Discussion for the most part deals with the works of Cochran and Hopkins [1961], Hills [1966], Lackenbruch [1965], Lackenbruch and Mickey [1968], Glick [1972, 1973], and Goldstein and Wolf [1977], all of which relate to the problem of estimation of error rates and the bias problem.

In keeping with the notation of Chapter 2, recall that we are considering two population groups Π_1 and Π_2 characterized by a sample space $\mathfrak{X}$ of s points, patterns, or states generated by a discrete vector $\mathbf{X} = (X_1, X_2, \ldots, X_p)$ with conditional (given Π_i) discrete mass functions $f_i, i = 1, 2$. If for the moment we assume that Π_1 and Π_2 are mixed with prior probabilities δ_1 and δ_2, then the unconditional density at some point $\mathbf{x}$ is given by

$$g(\mathbf{x}) = \delta_1 f_1(\mathbf{x}) + \delta_2 f_2(\mathbf{x}) = g_1(\mathbf{x}) + g_2(\mathbf{x}).$$

Glick [1972] refers to $g_i(\mathbf{x})$ as a discriminant score. A classification rule is an ordered partition $D = \langle D_1, D_2 \rangle$ of $\mathfrak{X}$ having the property that any randomly drawn point $\mathbf{x}$ is allocated to Π_i if and only if $\mathbf{x} \in D_i$. Given that $\mathbf{x} \in D_i$ the conditional probability of misclassification is given by $t(D|\mathbf{x}) = g_j(\mathbf{x})/g(\mathbf{x})$ for $i \neq j$. Therefore, the unconditional probability of misclassification is

$$t(D) = E\{t(D|\mathbf{X})\} = \sum_{D_1} g_2(\mathbf{x}) + \sum_{D_2} g_1(\mathbf{x}). \qquad (3.1\text{-}1)$$

A rule is optimal if it minimizes the unconditional probability of misclassification. That is, regard t as a function on the domain $\mathfrak{D}$ of all rules, and define

$$t^* = \inf_{D \in \mathcal{D}} t(D) \tag{3.1-2}$$

then D' is optimal if $t(D') = t^*$. Optimal rules can always be constructed by minimizing the conditional probability of misclassification at every point in $\mathcal{X}$. One specific partition D^* attributed to Welch [1939] and extended by Hoel and Peterson [1949] has regions defined by

$$D_1^* = \left\{ \mathbf{x} \mid g_1(\mathbf{x}) > g_2(\mathbf{x}) \right\}$$

$$D_2^* = \left\{ \mathbf{x} \mid g_1(\mathbf{x}) < g_2(\mathbf{x}) \right\} \tag{3.1-3}$$

points with equal discriminant scores are randomly assigned. It follows now that

$$t^* = t(D^*) = \sum_{\mathbf{x}} \min(g_1(\mathbf{x}), g_2(\mathbf{x})). \tag{3.1-4}$$

A trivial upper bound on t^* is $\frac{1}{2}$ since $\min(g_1, g_2) \leqslant \frac{1}{2} g \leqslant \frac{1}{2}$. Values of t^* close to $1/2$ indicate similar discriminant scores.

3.2 ESTIMATES OF THE OPTIMAL ERROR t^*

If prior probabilities and conditional discrete mass functions are not specified, then $t(D)$ cannot be evaluated for any arbitrary classification rule; in particular, $t(D^*)$ cannot be calculated. Suppose, however, that N individuals have been randomly sampled from the mixed population. The number from Π_i with $\mathbf{X} = \mathbf{x}$ is a binomial random variable $N_i(\mathbf{x})$ with expected value $N\delta_i f_i(\mathbf{x})$ for $i = 1$ or 2. Further, the total numbers of sampled individuals from the two groups are binomial random variables $N_1 = \Sigma N_1(\mathbf{x})$ and $N_2 = \Sigma N_2(\mathbf{x})$ with $N = N_1 + N_2$. Prior probabilities, conditional mass functions, and discriminant scores can now simply be estimated by $\hat{\delta}_i = (N_i/N)$, $\hat{f}_i(\mathbf{x}) = N_i(\mathbf{x})/N_i$, and $\hat{g}_i(\mathbf{x}) = N_i(\mathbf{x})/N$, for $i = 1$ or 2, respectively.

The plug-in estimate $\hat{t}$ of the unconditional probability of misclassification has values

$$\hat{t}(D) = \sum_{D_1} \hat{g}_2(\mathbf{x}) + \sum_{D_2} \hat{g}_1(\mathbf{x}) \tag{3.2-1}$$

for each rule $D \in \mathcal{D}$. Parroting the criterion of optimality, it is reasonable to use a sample-based classification rule that minimizes $\hat{t}$. Using the same argument as in Hoel and Peterson shows that $\hat{t}$ is minimized by the sample-based rule $\hat{D}$ defined by

$$\hat{D}_1 = \{\mathbf{x} | N_1(\mathbf{x}) > N_2(\mathbf{x})\}$$

$$\hat{D}_2 = \{\mathbf{x} | N_1(\mathbf{x}) < N_2(\mathbf{x})\} \tag{3.2-2}$$

and all points $\mathbf{x}$ such that $N_1(\mathbf{x}) = N_2(\mathbf{x})$ are randomly assigned. Hence it follows that

$$\inf_{D \in \mathcal{D}} \hat{t}(D) = \hat{t}(\hat{D}) = \sum_{\mathbf{x}} \frac{1}{N} \min(N_1(\mathbf{x}), N_2(\mathbf{x})). \tag{3.2-3}$$

The sample-based rule $\hat{D}$ generates two error-rates, namely, $\hat{t}(\hat{D})$ and $t(\hat{D})$. Using the terminology of Hills [1966] $\hat{t}(\hat{D})$ is called the "apparent" error, whereas $t(\hat{D})$ is referred to as the "actual" error. As Glick [1972] states, it is not clear whether $\hat{t}(\hat{D})$ is an estimate of $t(\hat{D})$ or $t(D^*)$. In terms of the relative size of the optimal, actual, and apparent error, it is immediate that

$$t(D^*) = \inf_{D \in \mathcal{D}} t(D) \leqslant t(\hat{D}). \tag{3.2-4}$$

Further, proposition A in Glick [1972] (which is also implicit in Hills [1966]) states that in general if $\hat{f}_i(\mathbf{x})$ is an unbiased estimate of $f_i(\mathbf{x}), i = 1, 2$, then the sample-based partition $\hat{D}$ induced by these estimates results in

$$E\{\hat{t}(\hat{D})\} \leqslant t(D^*) \leqslant t(\hat{D}). \tag{3.2-5}$$

Hence the mean apparent error $E\{\hat{t}(\hat{D})\}$ is an optimistically biased estimate of $t(D^*)$. Interestingly then, unbiasedness of density estimates used in the formulation of a classification rule leads to an unfavorable property with respect to estimating errors of misclassification. Note that the inequality given in (3.2-5) is satisfied when the sample-based partition (3.2-2) is used since $E\{N_i(\mathbf{x}) | N_i\} = f_i(\mathbf{x})$, for $i = 1$ or 2.

3.3   ESTIMATING THE MEAN ACTUAL ERROR

Lachenbruch [1965] has proposed a procedure to estimate the mean actual error of any given classification rule. The method is quite simple in that $N_1 - 1 + N_2$ sample points are used to classify the sample point omitted with this repeated for all objects from $\Pi_i, i = 1$ and 2. The proportions M_i / N_i misallocated are then recorded for both groups.

Recall that for the sample-based partition $\hat{D} = <\hat{D}_1, \hat{D}_2>$ the actual error $t(\hat{D})$ equals

$$\sum_{\hat{D}_1} g_1(\mathbf{x}) + \sum_{\hat{D}_2} g_2(\mathbf{x}) \qquad (3.3\text{-}1)$$

If we denote by

$$t_{N_1 - 1 + N_2}(\hat{D}_1)$$

the part of $\sum_{\hat{D}_1} g_1(\mathbf{x})$ that results from having used $(N_1 - 1) + N_2$ sample points, then $\delta_1 M_1 / N_1$ is an unbiased estimate of

$$E\left\{t_{N_1 - 1 + N_2}(\hat{D}_1)\right\}$$

and, unless N_1 is small, it differs only slightly from $E\{\sum_{\hat{D}_1} g_1(\mathbf{x})\}$. The same process is repeated for the observations from Π_2 with Lachenbruch's estimate assuming the form

$$\frac{\delta_1 M_1}{N_1} + \frac{\delta_2 M_2}{N_2} \qquad (3.3\text{-}2)$$

In the case where the underlying mass functions are multinomial, suppose for the moment we define each group Π_i on the basis of its parameter sets $\{p_{1j}\}, \{p_{2j}\}, j = 1, 2, \ldots, s$. Maximum likelihood estimates of these parameters are given by $\hat{p}_{ij} = (N_{ij}/N_i), i = 1, 2; \ j = 1, 2, \ldots, s$. Lachenbruch's estimate for the mean actual error in the multinomial case is rather easy to compute since it is only necessary to consider those states for which the allocation rule changes when one individual is removed.

Following Hills [1966], consider the sample of N_1 from Π_1 of whom M_1 are misallocated by the rule $\hat{D}$. If each sample member is now removed and allocated by the rule based on the remaining $N_1 - 1 + N_2$ sample points, the resulting proportion of misallocations will be, say, M_1'/N_1. Suppose $N_1 = N_2$ and $\Delta_j = \hat{p}_{1j} - \hat{p}_{2j}$ with s_1 of the s states having $\Delta_j = 0$ or $(1/N_1)$ and s_2 of the states having $0 < \Delta_j < (1/N_1)$. Then assuming equal prior probabilities, that is, $\delta_1 = \delta_2 = 1/2$

$$\frac{M_1'}{N_1} = \frac{M_1}{N_1} + \sum s_1 \hat{p}_{1j} + \frac{1}{2} \sum s_2 \hat{p}_{1j}. \qquad (3.3\text{-}3)$$

If $N_1 \neq N_2$ and one is not an exact multiple of the other and if s_1 states have $0 < \Delta_j < 1/N_1$, then

$$\frac{M_1'}{N_1} = \frac{M_1}{N_1} + \sum s_1 \hat{p}_{1j}. \qquad (3.3\text{-}4)$$

Hence to adjust the proportion M_1/N_1 and M_2/N_2 in the case of unequal sample sizes it is only necessary to identify those states for which $(-1/N_2)$ $<\Delta_j<(1/N_1)$, and to add the $\hat{p}_{1j}$ values for all states with $0<\Delta_j<(1/N_1)$ to (M_1/N_1) and the $\hat{p}_{2j}$ values for all states with $(-1/N_2)<\Delta_j<0$ to (M_2/N_2). States which are sparse are not counted since $\hat{p}_{1j},\hat{p}_{2j}$ will be zero.

Lachenbruch's estimate is sometimes inaccurately referred to as a "jackknife estimate." Following a suggestion made by Miller [1974], the actual and the apparent errors being biased estimates of t^* are logical candidates for the jackknife method. In its most simplistic form let $\hat{\theta}$ be an estimate of a parameter θ based on a sample of size N and $\hat{\theta}_{-j}$, the corresponding estimator based on a sample of size $N-1$ formed by deleting the jth observation. Define

$$\tilde{\theta}_j = N\hat{\theta} - (N-1)\hat{\theta}_{-j}, j=1,2,\ldots,N. \qquad (3.3\text{-}5)$$

Then the estimator

$$\tilde{\theta} = \frac{1}{N}\sum_{j=1}^{N}\tilde{\theta}_j = N\hat{\theta} - \frac{(N-1)}{N}\sum_{j=1}^{N}\hat{\theta}_{-j} \qquad (3.3\text{-}6)$$

has the property that it eliminates the order $1/N$ term from a bias of the form

$$E(\hat{\theta}) = \theta + \frac{a_1}{N} + \theta\frac{1}{N^2}. \qquad (3.3\text{-}7)$$

If $\hat{\theta}$ represents the actual error $t(\hat{D})$ based on a sample of size $N = N_1 + N_2$ from the mixed population, then the jackknifed actual error is given by

$$t(\hat{D}) = Nt(\hat{D}) - \frac{N-1}{N}\sum_{j=1}^{N}t_{-j}(\hat{D}) \qquad (3.3\text{-}8)$$

where $t_{-j}(\hat{D})$ is the actual error based on $N_1 + N_2 - 1$ observations. As in Lachenbruch's estimate, in the case of multinomial distributions the calculation is most straightforward. If in (3.3-8) both t and t_{-j} are replaced by $\hat{t}$ and $\hat{t}_{-j}$, then what results is the jackknifed apparent error. It would be interesting to see a study that investigates the properties of both jackknifed error rates with particular attention focused on the bias problem.

Both Cochran and Hopkins [1961] and Hills [1966] considered normal approximation arguments in determining estimates of the actual error.

Denoting the standard normal density by

$$h(t) = \frac{1}{\sqrt{2\Pi}} \exp\left(-\frac{1}{2}t^2\right)$$

and its cumulative by

$$H(t) = \int_{-\infty}^{t} h(x)dx$$

we suppose that the distribution of the difference $\hat{p}_{1j} - \hat{p}_{2j}$ is approximately normally distributed with mean $p_{1j} - p_{2j}$ and variance $\sigma_j^2 = \sigma_{1j}^2 + \sigma_{2j}^2$, where $\sigma_{ij}^2 = p_{ij}(1 - p_{ij})/N_i$, $i = 1, 2, j = 1, 2, \ldots, s$.

Hills showed that the mean value of M_1/N_1 can be approximated by

$$E\left(\frac{M_1}{N_1}\right) \cong \Sigma \left\{ p_{1j} G(-\lambda_j) - \frac{\sigma_{1j}^2}{\sigma_j} g(\lambda_j) \right\} \qquad (3.3\text{-}9)$$

where $\lambda_j = (p_{1j} - p_{2j})/\sigma_j$. Cochran and Hopkins considered the following estimate of $\Sigma_{\hat{D}_2} f_1(\mathbf{x})$

$$\frac{M}{N_1} + \Sigma \left(\frac{\hat{\sigma}_{ij}^2}{\hat{\sigma}_j}\right) g(\hat{\lambda}_j) \qquad (3.3\text{-}10)$$

where $\hat{\lambda}_j = (\hat{p}_{1j} - \hat{p}_{2j})/\sigma_j$. Recall that $\delta_1 \Sigma_{\hat{D}_2} f_1(\mathbf{x}) = \Sigma_{\hat{D}_2} g_1(\mathbf{x})$. The expectation of this estimate can be shown to be approximately

$$\Sigma_j \left\{ p_{1j} G(-\lambda_j) - \left(\frac{\sigma_{1j}^2}{\sigma_j}\right) g(\lambda_j) + \left(\frac{\sigma_{1j}^2}{\sigma_j}\right) 2^{-1/2} g\left(2^{-1/2}\lambda_j\right) \right\}. \qquad (3.3\text{-}11)$$

A further approach uses the approximation

$$E \underset{\hat{D}_2}{\Sigma} f_1(\mathbf{x}) \sim \Sigma p_{1j} G(-\lambda_j) \qquad (3.3\text{-}12)$$

with maximum likelihood estimates $\hat{p}_{ij}$, $i = 1, 2, j = 1, 2, \ldots, s$, replacing the unknown parameters. Hills shows that

$$E \Sigma \hat{p}_{1j} G(-\hat{\lambda}_j) \sim \left(\frac{-2^{-1/2}\sigma_{1j}^2}{\sigma_j}\right) g\left(2^{-1/2}\lambda_j\right) + p_{1j} G\left(-2^{-1/2}\lambda_j\right). \qquad (3.3\text{-}13)$$

TABLE 3.3-1

Mean Values of Actual and Apparent Errors of Allocation for Members of Π_1 Together with Expected Values of Two Estimates of Mean Actual Error for Three Multinomial Examples (Exact Values Given for First Two Rows of Each Example, Approximate Values for the Rest)

	SAMPLE SIZE $N_1 = N_2$	MEAN ACTUAL ERROR $E\sum f_1(\mathbf{x})$ $\hat{D}_2$	MEAN APPARENT ERROR $E(M_1/N_1)$	EXPECTED VALUE OF ESTIMATES OF THE MEAN ACTUAL ERROR COCHRAN	MAXIMUM LIKELIHOOD
Example 1	8	0.3815	0.2262	0.3331	0.2714
6 states	20	0.3432	0.2708	0.3335	0.2998
$\sum_{D_2} f_1(\mathbf{x}) = 0.3085$	50	0.3223	0.2917	0.3214	0.3064
$\sum_{D_1} f_2(\mathbf{x}) = 0.3085$	100	0.3168	0.2997	0.3150	0.3070
Example 2	8	0.2041	0.1148	0.1829	0.1482
6 states	20	0.1768	0.1377	0.1722	0.1540
$\sum_{D_2} f_1(\mathbf{x}) = 0.1587$	50	0.1687	0.1493	0.1659	0.1571
$\sum_{D_1} f_2(\mathbf{x}) = 0.1587$	100	0.1655	0.1545	0.1642	0.1590
Example 3	8	0.3401	0.0625	0.2120	0.1507
32 states	20	0.2538	0.0951	0.2067	0.1584
$\sum_{D_2} f_1(\mathbf{x}) = 0.1464$	50	0.1955	0.1038	0.1901	0.1469
$\sum_{D_1} f_2(\mathbf{x}) = 0.1536$	100	0.1722	0.1257	0.1716	0.1497
	150	0.1638	0.1330	0.1639	0.1494
	200	0.1596	0.1368	0.1600	0.1489

Source. Hills [1966].

57

This last estimate is referred to as the *maximum likelihood approximation estimate*.

For illustrative purposes, Hills considered two 6-state examples and one 32-state example in comparing the relative size of the true error rates and the corresponding components of the mean actual and mean apparent errors. In addition, he considered the expected values of the Cochran and maximum-likelihood estimates. The results of the various calculations are given in Table 3.3-1. As can clearly be seen from the table, the Cochran estimate approximates the mean actual error (really the first component of the mean actual error) more closely than the maximum likelihood estimate. Interestingly, however, the maximum-likelihood estimate in almost all cases comes closer to the true error than the mean actual error. Further, as we have noted before, the mean apparent error is optimistically biased, whereas the mean actual is pessimistically biased.

3.4 ASYMPTOTIC OPTIMALITY AND RATES OF CONVERGENCE

A number of questions immediately arise with respect to the large sample behavior of the actual and apparent errors. In particular, under what conditions do the two errors converge to the Bayes error, and if they do converge in what mode and at what rate? How pessimistically biased is the mean actual error and how optimistically biased is the mean apparent error? Glick's two fundamental papers adequately answer most of these questions. In his 1972 paper the uniform consistency of the sample-based error function $\hat{t}$ to the true error rate function is established. Glick's theorem is general and goes beyond the structure that we have imposed, namely, finite discrete sample spaces.

Theorem A (Glick [1972]). If the density estimators $\hat{f}_i, i = 1, 2$, each satisfies $0 \leqslant \hat{f}_i(\mathbf{x}) \to f_i(\mathbf{x})$ (with probability one) whenever $\delta_i > 0$, so that $0 \leqslant \hat{g}_i(\mathbf{x}) \to g_i(\mathbf{x})$ (with probability one), and if $\int \hat{\phi} \to 1$ (with probability one) where $\hat{\phi}(\mathbf{x}) = \hat{\delta}_1 \hat{f}_1(\mathbf{x}) + \hat{\delta}_2 \hat{f}_2(\mathbf{x})$, then

$$\sup_{D} |\hat{t}(D) - t(D)| \to 0 \text{ (with probability one)}.$$

Note that if the estimators $\hat{f}_i$ are densities, then $\int \hat{\phi} = 1$, and the last condition of the theorem is automatically satisfied. In particular, under the structure being considered here

$$\int \hat{\phi} = \sum \hat{\phi}(\mathbf{x}) = \sum_{\mathbf{x}} \left(\frac{N_1(\mathbf{x})}{N} + \frac{N_2(\mathbf{x})}{N} \right) = 1. \qquad (3.4\text{-}1)$$

Further, the theorem also remains true if convergence with probability one is replaced by convergence in probability.

In determining in what sense a sample-based classification rule is asymptotically optimal, Van Ryzin [1966] introduced the idea of a sequence of rules being Bayes-risk consistent.

Definition. Let $\overline{D}$ denote any sample-based classification rule (or more accurately, an element in a sequence of such rules constructed as sample size $N \to \infty$). The rule $\overline{D}$ is said to be Bayes-risk consistent if the error rate $t(\overline{D}) \to t^*$ in probability or in probability one.

Using this definition of asymptotic optimality and using Glick's theorems, it is a straightforward exercise to establish: (a) a density plug-in rule $\hat{D}$ is Bayes-risk consistent, that is, $t(\hat{D}) \to t^*$ (with probability one), or, the actual error converges (with probability one) to the optimal error and (b) the apparent error $\hat{t}(\hat{D}) \to t^*$ (with probability one). Note that the theorem immediately implies $|\hat{t}(\hat{D}) - t(\hat{D})| \to 0$; that is, the apparent and actual errors converge to the same number. Further

$$|\hat{t}(\hat{D}) - t^*| = \left| \inf_{D} \hat{t}(D) - \inf_{D} t(D) \right|$$

$$\leqslant \sup_{D} |\hat{t}(D) - t(D)| \text{ or } \hat{t}(\hat{D}) \to t^* \quad (3.4\text{-}2)$$

which establishes (b). Lastly, (a) follows by noting

$$|t(\hat{D}) - t^*| \leqslant |t(\hat{D}) - \hat{t}(\hat{D})| + |\hat{t}(\hat{D}) - t^*|$$

$$\to 0 \text{ (with probability one).} \quad (3.4\text{-}3)$$

Complementing the asymptotic optimality of the rule $\hat{D}$ (i.e., $P\{t(\hat{D}) \to t^*\} = 1$) is the question of how rapidly $t(\hat{D})$ tends to t^*. Questions further arise with respect to the convergence rate of the expected actual error to the optimum error. Following Glick [1973], let $m \geqslant 1$ denote the number of points $\mathbf{x}$ with $g_1(\mathbf{x}) \neq g_2(\mathbf{x})$, and let $d = \inf | \sqrt{g_1(\mathbf{x})} - \sqrt{g_2(\mathbf{x})} |$ over the set of points with unequal discriminant scores. Since the sample space is finite, m is finite and $1 > d > 0$. The following theorem shows that convergence is at least exponential.

Theorem B (Glick [1973]). For the classification rule $\hat{D}$ based on a sample of size N

$$(i) \quad 0 \leqslant 1 - P\{t(\hat{D}) = t^*\} \leqslant \frac{1}{2} m(1 - d^2)^N, \quad (3.4\text{-}4)$$

$$(ii) \quad 0 \leqslant E\{t(\hat{D})\} - t^* \leqslant \left(\frac{1}{2} - t^* \right)(1 - d^2)^N. \quad (3.4\text{-}5)$$

Cochran and Hopkins found that for fixed N, the disparity $E\{t(\hat{D})\} - t^*$ is generally larger for high t^*. Bound (ii) reinforces their work. Note that in a sense the term $\inf|\sqrt{g_1(x)} - \sqrt{g_2(x)}|$ acts as a pseudo distance between discriminant scores so that high error probability associated with similar discriminant scores results in $(1-d^2)^N$ being close to unity. Last, as Glick states, the bounds given in his theorem may be quite loose. We see later that this is particularly true for a similar class of bounds relating to the apparent error.

As we have seen, the mean apparent error $E\{\hat{t}(\hat{D})\}$ is less than or equal to the optimum error; that is, $\hat{t}(\hat{D})$ is an optimistically biased estimate of t^*. Knowing as we do that $P\{\hat{t}(\hat{D}) \to t^*\} = 1$, it follows that

$$t^* - E\{\hat{t}(\hat{D})\} \to 0$$

as $N \to \infty$. The following theorem relates the magnitude of the optimistic bias $t^* - E\{\hat{t}(\hat{D})\}$ to the sample size N and to the discriminant scores.

Theorem C (Glick [1973]). For a sample of size N, an upper bound on the pointwise bias of the estimator $\hat{w} = \min\{\hat{g}_1, \hat{g}_2\}$ is given by

$$\text{(iii)} \quad 0 \leqslant w(x) - E\{\hat{w}(x)\} \leqslant \frac{1}{2} N^{-1/2} \left[1 - |\sqrt{g_1(x)} - \sqrt{g_2(x)}|^2\right]^N. \quad (3.4\text{-}6)$$

At a point x with $g_1(x) = g_2(x)$, the bias $w(x) - E\{\hat{w}(x)\}$ has a lower bound proportional to $N^{-1/2}$. If the sample space contains no such point, then an exponentially decreasing upper bound on the total bias of the apparent error $\hat{t}(\hat{D})$ is given by

$$\text{(iv)} \quad 0 \leqslant t^* - E\{\hat{t}(\hat{D})\} \leqslant \frac{1}{2} mN^{-1/2}(1-d^2)^N. \quad (3.4\text{-}7)$$

The bound given in (3.4-7) is again consistent with the work of Cochran and Hopkins since they argued the bias to be $0(N^{-1/2})$, provided there exists one state where $g_1(x) = g_2(x)$. The theorem shows that in general large bias occurs with similar discriminant scores. We have more to say about the behavior of the apparent error in the next section. However, the bound given in (3.4-7), as was the case of the bound given in (3.4-5), tends to be quite loose, so much so that the convergence rate might be more rapid than $N^{-1/2}\alpha^N$ for $0 < \alpha < 1$. We conclude this section with Tables 3.4-1 and 3.4-2, adopted from Glick [1973], so that the magnitude of the upper bounds given in (3.4-5) and (3.4-7) for various sample sizes and values of d are illustrated.

TABLE 3.4-1

$(1 - d^2)^N$ UPPER BOUNDS

N	\multicolumn{4}{c}{d}
	1/4	1/16	1/64	1/256
1	0.9375	0.9961	0.9998	0.9999+
2	0.8789	0.9922	0.9995	0.9999+
4	0.7725	0.9845	0.9990	0.9999
8	0.5967	0.9692	0.9980	0.9999
16	0.3561	0.9393	0.9961	0.9998
32	0.1268	0.8823	0.9922	0.9995
64	0.0161	0.7784	0.9845	0.9990
128	0.0003	0.6059	0.9692	0.9980
256		0.3672	0.9394	0.9961
512		0.1348	0.8825	0.9922
1024		0.0182	0.7788	0.9845
2048		0.0003	0.6065	0.9692
4096			0.3678	0.9394
8192			0.1353	0.8825
16384			0.0182	0.7788

Source. Reproduced from N. Glick, "Sample-based Multinomial Classification," BIOMETRICS **29**:241–256, 1973, with permission of the Biometric Society.

TABLE 3.4-2

$N(1 - D^2)^N$ UPPER BOUNDS

N	\multicolumn{5}{c}{d}
	1/4	1/16	1/64	1/256	1/1024
1	0.9375	0.9961	0.9998	0.9999+	0.9999+
2	0.6215	0.7016	0.7068	0.7071	0.7071
4	0.3862	0.4922	0.4995	0.5000	0.5000
8	0.2110	0.3427	0.3529	0.3535	0.3535
16	0.0890	0.2348	0.2490	0.2499	0.2500
32	0.0224	0.1650	0.1754	0.1768	0.1768
64	0.0020	0.0973	0.1231	0.1249	0.1250
128		0.0536	0.0857	0.0882	0.0884
256		0.0229	0.0587	0.0623	0.0625
512		0.0060	0.0390	0.0439	0.0442
1024		0.0006	0.0243	0.0308	0.0312
2048			0.0134	0.0214	0.0221
4096			0.0057	0.0147	0.0156
8192			0.0015	0.0098	0.0110
16384			0.0001	0.0061	0.0077

Source. Reproduced from N. Glick, "Sample-based Multinomial Classification," BIOMETRICS **29**: 241–256, 1973, with permission of the Biometric Society.

3.5   THE EXACT BIAS OF THE APPARENT ERROR

Return now to the case in which the underlying mass functions from Π_1 and Π_2 are s state multinomials with parameter sets $\{p_{1j}\}$ and $\{p_{2j}\}$, respectively. Suppose instead of sampling from the mixture of the two subpopulations, independent random samples of size n and m are available from Π_1 and Π_2, respectively. Maximum likelihood estimates for the state probabilities in each group are then given by

$$\hat{p}_{1j} = \frac{n_j}{n}, \hat{p}_{2j} = \frac{m_j}{m}, \tag{3.5-1}$$

$j = 1, 2, \ldots, s$, and under the assumption of equal prior probabilities, the bias of the apparent error rate can be written as

$$\tfrac{1}{2} \sum_{j=1}^{s} \left[\min(p_{1j}, p_{2j}) - E \min(\hat{p}_{1j}, \hat{p}_{2j}) \right]. \tag{3.5-2}$$

Evaluation of the bias would require the calculation of $E \min(\hat{p}_{1j}, \hat{p}_{2j})$. Toward this end, suppose we consider a more general problem. Following Goldstein and Wolf [1977], let X and Y be independent integer-valued random variables assuming the values $0, 1, \ldots, n$ and $0, 1, \ldots, m$, respectively. Define $Z = \min(X/n, Y/m)$. Then, the probability distribution of Z is given by

$$P\{Z = z\} = \sum_{x > nz} P\{X = x\} P\{Y = mz\} I_{\left\{0, \frac{1}{m}, \frac{2}{m}, \ldots, 1\right\}}(z)$$

$$+ P\{X = nz\} P\{Y \geqslant mz\} I_{\left\{0, \frac{1}{n}, \frac{2}{n}, \ldots, 1\right\}}(z), \tag{3.5-3}$$

where $I_A(z)$ is the indicator function of the event A. It follows now that the expectation of Z is given by

$$E(Z) = \sum_{z=0}^{m-1} \sum_{x = \left[\left(\frac{n}{m}\right)z + 1\right]} \left(\frac{z}{m}\right) P\{X = x\} P\{Y = y\}$$

$$+ \sum_{z=0}^{m} \sum_{y > \left[\left(\frac{m}{n}\right)z\right]} \left(\frac{z}{n}\right) P\{X = x\} P\{Y = y\} \tag{3.5-4}$$

where $[(n/m)z + 1]$ is the greatest integer less than or equal to $(n/m)z + 1$. Since $n\hat{p}_{1j}$ and $m\hat{p}_{2j}$ are independent binomial random variables with parameters (n, p_{1j}) and (m, p_{2j}), respectively, it follows that the bias of the apparent error is equal to

$$
\frac{1}{2} \sum_{j=1}^{s} \left[\min(p_{1j}, p_{2j}) - \left\{ \sum_{z=0}^{m-1} \sum_{x=\left[\left(\frac{n}{m} \right) z + 1 \right]}^{n} (z/m) b(n, p_{1j}; x) b(m, p_{2j}; z) \right.\right.
$$

$$
\left.\left. + \sum_{z=0}^{n} \sum_{y > \left[\left(\frac{m}{n} \right) z \right]}^{m} \left(\frac{z}{n} \right) b(n, p_{1j}; z) b(m, p_{2j}, y) \right\} \right] \qquad (3.5\text{-}5)
$$

where $b(n, p, r) = \binom{n}{r} p^r (1-q)^{n-r}, \ 0 < p < 1, \ q = 1 - p, \ r = 0, 1, 2, \ldots, n$.

Table 3.5-1 presents the results of evaluating the exact bias given in (3.5-5) for various combinations of state probabilities p_{1j} and p_{2j} and for equal state sample sizes m and n. The evaluation of the exact bias reinforces the work of Glick and Cochran and Hopkins in this area and shows some additional properties of the apparent error. In particular:

1. When the sample sizes are held fixed and the difference between p_{1j} and p_{2j} increases, then the bias decreases. This property also holds for unequal sample sizes.
2. When the difference between p_{1j} and p_{2j} is fixed and $n + m$ increases, then the bias decreases. Further, when the samples are fixed and both p_{1j} and p_{2j} increase, but with $p_{1j} - p_{2j}$ remaining constant, the bias increases.
3. For any fixed value of $m + n$ (divisible by 2) the bias is least (all other things remaining equal) when $m = n$. Hence the values reported in Table 3.5-1 represent the best possible results among all splits of $m + n$ into two groups.

A question remaining unanswered from Glick's [1973] paper concerned the practicality of the derived bounds. As he stated, the bounds may indeed be quite loose and hence of limited use in an applied problem where the researcher is interested in getting a reading on the bias through only an evaluation of the apparent error. In investigating this issue we consider a tighter bound on the bias for the apparent error that is implicit in Glick's work, namely:

$$
\frac{1}{2} (m+n)^{-1/2} \sum_{j=1}^{s} \left[1 - \left\{ (\delta p_{1j})^{1/2} - ((1-\delta) p_{2j})^{1/2} \right\}^2 \right]^{m+n}. \qquad (3.5\text{-}6)
$$

TABLE 3.5-1
EXACT BIAS OF APPARENT ERROR RATE

$n = m$	p_j	0.025	0.05	$p_{1j} - p_{2j}$ 0.075	0.10	0.15	0.20
5	0.05	0.0176	0.0156	0.0138	0.0121	0.0093	0.0070
5	0.10	0.0291	0.0259	0.0230	0.0203	0.0157	0.0120
5	0.15	0.0370	0.0331	0.0295	0.0263	0.0205	0.0158
5	0.20	0.0429	0.0385	0.0345	0.0308	0.0242	0.0187
7	0.05	0.0158	0.0135	0.0115	0.0097	0.0070	0.0050
7	0.10	0.0250	0.0215	0.0185	0.0158	0.0113	0.0080
7	0.15	0.0314	0.0273	0.0236	0.0204	0.0149	0.0107
7	0.20	0.0360	0.0316	0.0276	0.0239	0.0177	0.0128
10	0.05	0.0134	0.0108	0.0087	0.0069	0.0043	0.0027
10	0.10	0.0208	0.0172	0.0141	0.0115	0.0075	0.0047
10	0.15	0.0259	0.0217	0.0181	0.0150	0.0101	0.0066
10	0.20	0.0297	0.0252	0.0212	0.0178	0.0122	0.0081
15	0.05	0.0108	0.0080	0.0059	0.0043	0.0023	0.0012
15	0.10	0.0165	0.0128	0.0099	0.0076	0.0043	0.0023
15	0.15	0.0205	0.0164	0.0129	0.0101	0.0060	0.0034
15	0.20	0.0235	0.0190	0.0153	0.0122	0.0074	0.0043
20	0.05	0.0090	0.0063	0.0043	0.0029	0.0013	0.0005
20	0.10	0.0138	0.0102	0.0074	0.0054	0.0027	0.0013
20	0.15	0.0172	0.0131	0.0099	0.0073	0.0038	0.0019
20	0.20	0.0197	0.0154	0.0118	0.0090	0.0049	0.0025
30	0.05	0.0068	0.0042	0.0025	0.0015	0.0005	0.0002
30	0.10	0.0105	0.0071	0.0047	0.0030	0.0011	0.0004
30	0.15	0.0132	0.0093	0.0065	0.0044	0.0018	0.0007
30	0.20	0.0152	0.0111	0.0079	0.0055	0.0025	0.0010
50	0.05	0.0045	0.0022	0.0011	0.0005	0.0001	0.0000
50	0.10	0.0072	0.0042	0.0023	0.0012	0.0003	0.0001
50	0.15	0.0092	0.0057	0.0034	0.0019	0.0005	0.0001
50	0.20	0.0107	0.0070	0.0043	0.0026	0.0008	0.0002

Source. Reproduced from M. Goldstein and E. Wolf, "On the Problem of Bias in Multinomial Classification," BIOMETRICS **33**: 328–329, 1977, with permission of the Biometric Society.

Before proceeding, however, we want to remind the reader that in all of Glick's work nonerror rates are used, whereas error rates are used here. However, when looking at differences, as we are here, the same results follow provided no state yields the same discriminant scores. Further, in deriving the exact bias independent samples are assumed, whereas in Glick's work sampling from a mixture of two subpopulations is assumed. However, again the same results prevail, the only difference being in the way parameters are estimated.

Table 3.5-2 presents three sample problems containing 4, 8, and 16 states with equal sample sizes of 50, 100, and 200, respectively. As is most evident, the disparity between the exact bias and the tightened version (3.5-6) of Glick's bound is quite large over all three examples. Hence it appears that some other approach is needed in assessing the bias of the apparent error. This is a most important problem in discriminant analysis and, in particular, in discrete discriminant analysis.

TABLE 3.5-2

A Four-, Eight-, and 16-State Problem

	a			b			c	
p_{1j}	p_{2j}	EXACT STATE BIAS	p_{1j}	p_{2j}	EXACT STATE BIAS	p_{1j}	p_{2j}	EXACT STATE BIAS
0.30	0.035	0.00875	0.175	0.20	0.00584	0.075	0.125	0.00028
0.025	0.020	0.00697	0.15	0.175	0.00529	0.10	0.15	0.00046
0.20	0.25	0.00697	0.10	0.125	0.00397	0.03	0.04	0.00166
0.25	0.20	0.00697	0.20	0.15	0.00251	0.125	0.10	0.00191
			0.125	0.15	0.00468	0.095	0.09	0.00460
			0.10	0.025	0.00006	0.075	0.025	0.00003
			0.075	0.125	0.00119	0.02	0.035	0.00075
			0.075	0.05	0.00225	0.05	0.06	0.00245
						0.10	0.075	0.00144
						0.04	0.025	0.00094
						0.05	0.085	0.00044
						0.075	0.05	0.00092
						0.05	0.06	0.00245
						0.025	0.03	0.00213
						0.055	0.025	0.00024
						0.035	0.025	0.00143
EXACT BIAS		0.02966			0.02579			0.02213
GLICK BOUND		0.176			0.200			0.251

Source. Reproduced from M. Goldstein and E. Wolf, "On the Problem of Bias in Multinomial Classificaton," BIOMETRICS **33**: 328–329, 1977, with permission of the Biometric Society.

Based on the material discussed in this chapter two simple approaches come to mind with respect to estimating the bias when using the apparent error as a measure of performance. One approach would replace the unknown discriminant scores g_1 and g_2 by their maximum-likelihood estimates in Glick's bound (3.4-7). Using this approach yields an estimated upper bound on the bias of the mean apparent error and would have the form

$$\frac{1}{2} mN^{-1/2} \left(1 - \left\{ \inf \left(\sqrt{N_1(\mathbf{x})/N} - \sqrt{N_2(\mathbf{x})/N} \right) \right\}^2 \right). \quad (3.5\text{-}7)$$

This approach, however, will probably not be too satisfying since the unestimated bound (3.4-7) is itself quite loose, and its estimate might be, for practical purposes, of very little value.

A further approach relies on the exact expression (3.5-5) developed in this section with unknown parameters replaced by maximum-likelihood estimates. An estimate for the bias then assumes the form

$$\sum_{j=1}^{s} \left\{ \min(\hat{p}_{1j}, \hat{p}_{2j}) - \hat{E}\min(\hat{p}_{1j}, \hat{p}_{ij}) \right\}, \quad (3.5\text{-}8)$$

where by $\hat{E}(\cdot)$ we mean the expectation resulting when using the probability distribution induced through the observed state frequencies. In the case of empty states some small positive number can be used in place of zero with the other state frequencies adjusted so that the usual constraints on the empirical mass functions are realized. This approach appears to be promising, and indeed some limited sampling experiments have indicated that the distribution of (3.5-8) under limited sparseness is rather symmetric about the true bias with a well-behaved variance. Further work using this approach is needed to determine whether it is indeed a viable and economical method.

3.6   AFTERTHOUGHT

Glick has shown that the bias of the actual error for the sample-based partition (3.2-2) converges to zero at least as rapidly as $\alpha^N (0 < \alpha < 1)$, whereas the bias for the apparent error converges at least as rapidly as $N^{-1/2}\alpha^N$, provided that there exists no state that emits equal discriminant scores. Indeed, at least in the latter case, Section 3.5 intimates that the convergence might be more rapid than the exponential rate given by Glick's bound. These convergence rates compare most favorably to equivalent results assuming underling multivariate normal distributions.

In the usual case of two multivariate normal distributions differing only in their mean vectors, the Bayes procedure is a linear rule, the sample-based version of which is usually referred to as the Fisher linear discriminant function. Okamoto [1963] established that under such a structure the actual error converges like $0(N^{-1})$. Similar work by Sorum [1971], Lachenbruch and Mickey [1968], and Dunn [1971] show through Monte Carlo work various convergence rates of $N^{-1/2}$. Recent work by McLachlan [1976] shows the apparent error for normal structures converging like $0(N^{-1})$.

The more favorable position of multinomial rules vis à vis convergence rates prompted Cochran and Hopkins to advise that under certain instances, intervals be discretized and some discrete rule utilized instead of blatant application of the Fisher linear rule.

CHAPTER 4

The Variable-Selection Problem

4.1 INTRODUCTION

As we have seen, a few variables, each assuming no more than two or three levels, can result in a rapid proliferation of states, which in situations of limited sample information can cause difficulty in estimation. This, coupled with the potential problems associated with missing data, has often resulted in practitioners shying away from utilizing true distributional properties that govern discrete measurements when forming discriminant functions. In Chapter 2 we presented some procedures that appear to better deal with the problem than, for example, the full multinomial; however, if all variables are maintained the issue still potentially remains. Hence it would appear then that methods for deriving "good" subsets of variables in the discrete case for use in forming a discriminant function are of great interest.

Assuming as a starting condition multivariate normal structures and the use of Fisher's linear discriminant function, a number of variable-selection procedures are in use. Weiner and Dunn [1966], for example, discuss three methods of variable selection:

1. Use of t-statistics. A simple and obvious statistic that could be used to rank the variables for entry is

$$t_k = \frac{\bar{x}_{1k} - \bar{x}_{2k}}{s_k \sqrt{\dfrac{1}{N_1} + \dfrac{1}{N_2}}}$$

where s_k^2 is the usual pooled estimate of the variance for variable X_k. On the basis of the computed t_k values the variables are ordered, and hence a "good" subset can be chosen. Note, however, that if a rule is to include all variables such that the overall significance level is α, a simultaneous procedure should be used. That is, the joint significance level must be taken into account since k variables are being tested.

2. Discriminant function coefficient. From the computed discriminant function we can rank the variables according to their standardized discriminant coefficients and select only those variables having "large" coefficients.

3. Stepwise selection. The discriminant function can be computed using a stepwise regression program that introduces variables one at a time. The criterion for entry at each step is to select that variable that reduces the residual sum of squares as much as possible.

Although all these procedures are to some extent ad hoc, their use can certainly be better justified assuming multivariate normality than if the underlying vectors were say dichotomous. Moreover, in the context of the variable-selection problem these two situations can be shown to have some interesting differences. For example, certain structural relationships between variables will lead to increased discriminatory power assuming normality, whereas a polar negative effect can result if the variables are dichotomous. In particular, when the distributions in two groups are multivariate normal with different mean vectors but the same covariance matrix, Cochran [1962] argued in favor of the two following points: (a) if all variables are independent, then the best set of k variables consists of the k best single variables and (b) for a set of two variables, any negative correlation reduces the probability of misclassification, whereas a positive correlation, unless it is sufficiently high, increases the probability of misclassification. This last point runs counter to the work of Elashoff, Elashoff, and Goldman [1967] where if multinomial distributions are assumed generated by p dichotomous random variables it is argued that positive correlation may decrease misclassification probability while negative correlation may decrease the same.

In the sections to follow we discuss some variable selection procedures discussed in the literature that all share the property of utilizing the underlying multinomial distributions assumed to be present. Goldstein and Rabinowitz [1975] utilize the bounds derived by Glick [1973] to define a "best" subset. In addition, three sequential procedures due to Raiffa [1961], Lachin [1973], and Goldstein and Dillon [1977] rank variables for entry, and on the basis of established thresholds either include a variable for entry or delete it from consideration. Two data bases illustrate the procedures discussed, with the exception of the Raiffa procedure, which unfortunately is not operational at this time.

4.2   A PROCEDURE BASED ON DIFFERENCES IN DISCRIMINANT SCORES

In Glick's work [1973] the expression $(1 - d^2)^N$, where

$$d = \inf{}_x \left| \sqrt{g_1(x)} - \sqrt{g_2(x)} \right|$$

is of fundamental importance. In some sense d is a measure of distance between the discriminant scores with, in general, large values of d associated with better discrimination. Goldstein and Rabinowitz [1975] use the pseudo distance d in defining a variable-selection procedure.

When a given classification rule based on the entire random p-dimensional vector X is replaced by using a certain subset of variables $X(h)$, d will change. Associating large values of d with better discrimination led the authors to consider a "best" subset as one satisfying

$$\max_{1 \leqslant j \leqslant p} \; \max_{1 \leqslant i \leqslant \binom{p}{j}} \; \min_{x(i)} \left| \sqrt{g_1(x(i))} - \sqrt{g_2(x(i))} \right| \qquad (4.2\text{-}1)$$

where $x(i)$ represents a particular combination when j out of the available p variables are used, $j = 1, 2, \ldots, p$; $i = 1, 2, \ldots, \binom{p}{j}$. A sample-based analogue to (4.2-1) is

$$\max_{1 \leqslant j \leqslant p} \; \max_{1 \leqslant h \leqslant \binom{p}{j}} \; \min_{x(h)} \left| \sqrt{N_1(x(h))} - \sqrt{N_2(x(h))} \right| \qquad (4.2\text{-}2)$$

where $N_i(x(h))$ is the number of sample observations from group Π_i, $i = 1$, 2 having $X(h) = x(h)$.

The expression given by (4.2-2) needs to be amended to account for the dimensionality of the sample space induced by restricting consideration to $X(h)$. When the number of variables used is increased, the number of points in the sample space increases multiplicitly, and hence for a fixed total sample of size N the expected number of observations at any point $x(h)$ decreases. We would expect, therefore, that the difference $|(N_1(x(h)))^{\frac{1}{2}} - (N_2(x(h)))^{\frac{1}{2}}|$ will be small when both $N_1(x(h))$ and $N_2(x(h))$ are expected to be small. Scaling the frequency estimates thus seems appropriate, and indeed the authors consider the following expression

$$\max_{1 \leqslant j \leqslant p} \; \max_{1 \leqslant h \leqslant \binom{p}{j}} \; \min_{x(h)} \left| \frac{\sqrt{N_1(x(h))} - \sqrt{N_2(x(h))}}{(\Pi_j l_j)^{\frac{1}{2}}} \right| \qquad (4.2\text{-}3)$$

where l_j is the number of levels of variable j and where the product is over those variables other than those in $X(h)$. In particular, if X is a multivariate binary vector of dimension p, then $\Pi_j l_j = 2^{p-k}$.

In applying (4.2-3) it is generally found that too many variables are selected. Because of this tendency a variation can be utilized, namely:

$$\max_{1 \leqslant j \leqslant p} \ \max_{1 \leqslant h \leqslant \binom{p}{j}} \ \operatorname*{avg}_{x(h)} \left| \frac{\sqrt{N_1(x(h))} - \sqrt{N_2(x(h))}}{(\Pi_j l_j)^{\frac{1}{2}}} \right|. \qquad (4.2\text{-}4)$$

In words, (4.2-4) chooses that subset of variables that maximizes the scaled average value of $|(N_1(x(h)))^{\frac{1}{2}} - (N_2(x(h)))^{\frac{1}{2}}|$.

4.3   FORWARD AND STEPWISE PROCEDURES

4.3.1   Raiffa's Methods

Raiffa [1961] has considered essentially two methods of variable selection when the underlying predictor set is representable as a p-dimensional dichotomous vector. The first method that we discuss is referred to by Raiffa as a *forward Bayes sequential procedure*. For consistency with his paper we change notation slightly; however, all terms introduced are identified when appropriate to what was discussed before.

Two populations, w_1 and w_2, are assumed to have mixing probabilities $\Pi^{(1)}$ and $\Pi^{(2)}$, respectively. For a given decision rule δ the probability of misclassifying an item from w_i is denoted by $\alpha_i(\delta)$, $i = 1, 2$. For the first time we introduce losses a_1 and a_2 due to wrong terminal decisions when w_1 and w_2, respectively, are true and a cost c_j associated with observing variable $X_j, j = 1, 2, \ldots, p$. Note that in our previous notation w_i was denoted by Π_i and $\Pi^{(i)}$ by δ_i. Further, the misclassification probabilities $\alpha_i^{(\delta)}$ were expressed as $\Sigma_{D_i} g_i(\mathbf{x})$, $i = 1, 2, j = 1, 2, i \neq j$.

Now let $\rho_0(\Pi^{(1)})$ be the Bayes risk against $\Pi^{(1)}$ when no variables are observed, namely

$$\rho_0(\Pi^{(1)}) = \max\left[a_2(1 - \Pi^{(1)}), a_2 \Pi^{(1)} \right]. \qquad (4.3\text{-}1)$$

If variable X_j is used and a Bayes rule is used against a prior distribution Π, then the resulting Bayes risk is given by

$$\rho_j(\Pi) = \min_{\delta \in D_j} \left[\Pi a_1 \alpha_1(\delta) + (1 - \Pi) a_2 \alpha_2(\delta) \right] + c_j \qquad (4.3\text{-}2)$$

where δ is a rule within the class of rules D_j that uses variable X_j. The forward Bayes selection procedure has the following form:

1. The process terminates without any items being chosen if

$$\rho_0(\Pi^{(1)}) \leqslant \min_{1 < j \leqslant p} \rho_j(\Pi^{(1)}). \tag{4.3-3}$$

2. If this does not happen, then use $X_{[1]}$ (where $[1]$ is one of the integers $1, 2, \ldots, p$) where

$$\rho_{[1]}(\Pi^{(1)}) = \min_{1 \leqslant j \leqslant p} \rho_j(\Pi^{(1)}). \tag{4.3-4}$$

3. Let $X_{[1]}, X_{[2]}, \ldots, X_{[k]}$ be the first k variables sequentially selected by the procedure, and suppose the values of these variables are given by $x_{[1]}, x_{[2]}, \ldots, x_{[k]}$. Define the prior probability of $w_1^{(k+1)}$ by

$$\Pi^{(k+1)} = P\left\{ w_1^{(k+1)} \right\} = P\left\{ w_1^{(1)} | X_{[1]} = x[1], \ldots, X_{[k]} = x_{[k]} \right\} \tag{4.3-5}$$

where state $w_i^{(k+1)}$ is $\{ X_{[1]} = x_{[1]}, \ldots, X_{[k]} = x_{[k]}, w_i \}$. To determine whether an additional variable $X_{[r]}$ is to be added to the set of k previously selected variables, we define

$$\rho_0^{(k+1)}(\Pi^{(k+1)}) = \max\left[a_2 (1 - \Pi^{(k+1)}), a_1 \Pi^{(k+1)} \right], \tag{4.3-6}$$

and let $\rho_r^{(k+1)}(\Pi^{(k+1)})$ be the Bayes risk including the cost of observing variable X_r, but not the costs of $X_{[1]}, X_{[2]}, \ldots, X_{[k]}$. The process terminates at stage k if

$$\rho_0^{(k+1)}(\Pi^{(k+1)}) \leqslant \min_{r \notin \{[1], \ldots, [k]\}} \rho_r^{(k+1)}(\Pi^{(k+1)}) \tag{4.3-7}$$

and chooses $X_{[k+1]}$ if

$$\rho_{k+1}^{(k+1)}(\Pi^{(k+1)}) = \min_{r \notin \{[1], \ldots, [k]\}} \rho_r^{(k+1)}(\Pi^{(k+1)}). \tag{4.3-8}$$

The second procedure that we discuss here is what Raiffa calls a *forward Bayes nonsequential procedure of order* $s = 1$. This method proceeds as follows: Let $\rho_{X_{I_1} \ldots X_{I_s}}$ be the Bayes risk when variables $X_{I_1}, X_{I_2}, \ldots, X_{I_s}$ are used where losses due to wrong terminal decisions as well as costs of variables employed are included.

At stage 1 let $X_{[1]}$ be the item for which $\rho_{[1]} \leqslant \rho_j, j = 1, 2, \ldots, p$; at stage 2 let $X_{[2]}$ be the item for which $\rho_{[1],[2]} \leqslant \rho_{[1],j}, j \neq [1], j = 1, 2, \ldots, p$; and at stage

$k+1$ let $I_{[k+1]}$ be the item for which

$$\rho_{[1],\ldots,[k],[k+1]} \leqslant \rho_{[1],\ldots,[k],j}, \qquad \begin{array}{l} j \neq [1],\ldots,[p] \\ j = 1,2,\ldots,p \end{array}.$$

The process terminates at stage l, where l is the smallest integer for which $\rho_{[1],\ldots,[l],[l+1]} > \rho_{[1],[2],\ldots,[l]}$.

As we have stated in Section 4.1, Raiffa's procedures are not computer-operational, and hence have been included for completeness but not for illustration with real data sets. The ideas that he sets forth, however, are interesting and potentially useful. The reader is advised to read Raiffa's paper for a more thorough discussion.

4.3.2 Lachin's Procedure

Lachin [1973] has proposed a stepwise selection procedure that is not restricted to dichotomous responses and that has the desirable property of deleting a variable at some later stage. In setting the stage for discussion of the procedure, we first introduce notation and some distributional results.

Suppose a sample space is generated by a $t+1$ dimensional discrete vector $(X_0, X_1, \ldots, X_t)$. If the number of levels of variable X_i is s_i, $i = 0, 1, \ldots, t$ the sample space consists of $s = \Pi_{i=0}^{t} s_i$ points. Assume now that a sample of size N is available and denote by $a_{ij \ldots k}$, the frequency of points having the property that $X_0 = i$, $X_1 = j, \ldots, X_t = k$. Lachin considers models of the form

$$a_{ij \ldots k} = N \rho_i \rho_{j \ldots k} \gamma_{ij \ldots k} \qquad (4.3\text{-}9)$$

where ρ_i represents the parametric marginal proportion of the ith category of variable X_0, $\rho_{j \ldots k}$ represents the portion for the $j \cdots k$ state of the joint distribution of the variables $X_1, X_2, \ldots, X_t$, and $\gamma_{ij \ldots k}$ represents the residual that is a measure of the error attained by assuming the particular set of parameters specified. Model (4.3-9) represents the hypothesis that the first variable X_0 is independent of the remaining variables $X_1, X_2, \ldots, X_t$.

If we assume that variable X_0 has two levels that we think of as the defining characteristic of the two populations of interest, then the sample may be viewed as comprising a contingency table of dimension $2 \times c$ where $c = \Pi_{j=1}^{t} s_j$. Lachin shows that if the usual maximum likelihood estimates, namely,

$$p_i = \frac{1}{N} \Sigma_{j \ldots k} a_{ij \ldots k}, \quad p_{ij \ldots k} = \frac{a_{ij \ldots k}}{N}$$

and

$$p_{j\cdots k} = \frac{1}{N}\Sigma_i a_{ij\cdots k}$$

are used, then if $q_{ij\cdots k} = \dfrac{p_{ij\cdots k}}{p_i}$, the expression

$$Q = \sum_{j\cdots k} \frac{q_{2j\cdots k} q_{1j\cdots k}}{p_{j\cdots k}} \tag{4.3-10}$$

results in $N(1-Q)$ being asymptotically χ^2, say, $\chi^2_{(1)}$ with $c-1$ degrees of freedom. Suppose an additional variable X_{t+1} is measured. The resulting table will now be of dimension $2 \times d$ where $d = cs_{t+1}$. The resulting statistic analogous to (4.3-10) with this additional variable will again be asymptotically χ^2, say, $\chi^2_{(2)}$ with degrees of freedom $d-1$. Lachin shows that in general $\chi^2_{(2)} \geqslant \chi^2_{(1)}$ with equality only when X_{t+1} is independent of $(X_0, X_1, \ldots, X_t)$. In addition, the statistic $\chi^2_{(3)} = \chi^2_{(2)} - \chi^2_{(1)}$ is also asymptotically χ^2 with degrees of freedom $d-c$. As Lachin states, $\chi^2_{(3)}$ is a statistic equivalent to what is used to test the hypothesis of the conditional independence of X_0 and X_{t+1} given the levels of the remaining variables $(X_1, X_2, \ldots, X_t)$ and, therefore, measures the change in the degree of association between X_0 and the remaining variables where X_{t+1} is added to the predictor set.

The stepwise procedure at any stage is dependent on the magnitude of the statistic $\chi^2_{(3)}$. In particular, utilization of Lachin's procedure first requires that two values be specified, to wit: (a) the minimum acceptable probability level for the addition of a variable (MPA) and (b) the minimum acceptable probability level for the deletion of a variable already selected (MPD). The only restriction is that MPA is equal to or greater than MPD.

Suppose now we assume that at some point in the process the set P of predictor variables contains m variables $[X_{I_1}, X_{I_2}, \ldots, X_{I_m}]$, where I_j is the jth variable selected, and that after the mth variable has been added the resulting χ^2 value is $\chi^2(m)$ with degrees of freedom $df(m)$. We now test for the deletion of each predictor variable in P by computing $\Delta\chi^2(I_j) = \chi^2(m) - \chi^2(I_j)$ with $\Delta df(I_j) = df(m) - df(I_j)$, where $\chi^2(I_j)$ and $df(I_j)$ are derived from the two-way table with X_{I_j} removed from the joint distribution of the set P. If probability $(\Delta\chi^2(I_j)) > \text{MPD}$ for any j, then X_{I_j} is deleted from the set P and the next step is begun. Otherwise, each of the remaining $t - m = n$ variables not in the set of predictors, say, $\overline{P} = (X_{J_1}, \ldots, X_{J_n})$, is examined for addition to P. Note that probability $(\Delta\chi^2(I_j)) > \text{MPD}$ means that there is not sufficient evidence in the data to reject the hypothesis of independence of X_0 and X_{I_j} in the presence of the remaining predictor variables to warrant its inclusion in P. The process now continues by computing for each variable in $\bar{P}$, $\Delta\chi^2(J_k) = \chi^2(J_k) - \chi^2(m)$, $\Delta df(J_k) = df(J_k) - df(m)$, where by $\chi^2(J_k)$ we mean the χ^2 value obtained by adding variable X_{J_k} to the predictor set and $df(J_k)$, the resulting degrees of freedom. Variables are then ranked according to how small the resulting probability ($\Delta\chi^2(J_k)$) with the smallest such probability denoted by p'. If $p' < \mathrm{MAP}$, then X_{J_k} is added to P and the next step begins. If not, the process terminates. If the data suffer from sparseness, Lachin suggests a slightly different approach in determining the appropriate degrees of freedom. For example, if in a sample of size N, $\rho_{j\ldots k} = 0$ for some $x \epsilon \mathcal{X}$, then the sample marginal space has $c' < c$ nonnull states. In such cases, Lachin proposes that c be used as the degrees of freedom instead of $c - 1$ (or $c' - 1$). If a χ^2 value is found by subtraction then its degrees of freedom are also found by subtraction.

Lachin's procedure is particularly appealing since computer routines are readily available for its use. Moreover, it has the nice property of being able to delete a variable after it has been entered. However, we see in the next procedure to be discussed that it suffers from an inability (at least in the way it is defined above) in not permitting additions or deletions on the basis of certain levels of a variable.

4.3.3   A Procedure Based on a Kullback Divergence Statistic

Goldstein and Dillon [1977] proposed a forward variable-selection procedure that relies on Kullback's work on information theory and extends work by Hills [1967]. The procedure allows for adding variables to given levels of variables that have already been made part of the system.

Suppose that the dichotomous random variables $Z_1, Z_2, \ldots, Z_p$ generate 2^p multinomial states within two disjoint groups G_1 and G_2. In each group G_i, the state probabilities are denoted by p_{ij}, $i = 1,2$; $j = 1,2,\ldots,s$, where $s = 2^p$. Following Kullback [1959], we define the mean information per observation from G_1 for discriminating in favor of G_1 against G_2 by

$$I(1:2) = \sum_{j=1}^{s} p_{1j} \log \frac{p_{1j}}{p_{2j}}. \qquad (4.3\text{-}11)$$

Information is additive in that a sample of N independent observations 0_N from the same population provides N times the mean information in a single observation, that is, $I(1:2; 0_N) = NI(1:2)$. Note now that

$$I(2:1) = \sum_{j=1}^{s} p_{2j} \log \frac{p_{2j}}{p_{1j}} \qquad (4.3\text{-}12)$$

and hence

$$I(1:2) + I(2:1) = J(1,2) = \sum_{j=1}^{s} (p_{1j} - p_{2j}) \log \frac{p_{1j}}{p_{2j}}. \qquad (4.3\text{-}13)$$

The expression $J(1,2)$ is a measure of the divergence between the two groups, with large values of J in an absolute sense indicating ease of discrimination.

Suppose now that two independent random samples of size N_1 and N_2 are taken from the s-state multinomial distribution one under the parameters of G_1 and the other under G_2. Let the random variables W_{ij} be defined as the number of observations from group i falling in state j for $i = 1,2$; $j = 1, 2, \ldots, s$, and suppose that the observed state frequencies are given by $(w_1) = (w_{11}, w_{12}, \ldots, w_{1s})$, $(w_2) = (w_{21}, w_{22}, \ldots, w_{2s})$, where $\Sigma w_{1j} = N_1$, $\Sigma w_{2j} = N_2$. Consider the following hypotheses:

H_1:   The samples are from different populations

$$(p_1) = (p_{11}, p_{12}, \ldots, p_{1s}), \; (p_2) = (p_{21}, p_{22}, \ldots, p_{2s})$$

H_2:   The samples are from the same population

$$(p_0) = (p_{01}, p_{02}, \ldots, p_{0s}), \; p_{1j} = p_{2j} = p_{0j}, j = 1, 2, \ldots, s$$

Since the samples are independent, it follows that the information in favor of H_1 as against H_2 is given by

$$I(H_1 : H_2) = \sum_{(w_1),(w_2)} p_1(w_1) p_2(w_2) \log \frac{p_1(w_1) p_2(w_2)}{p_0(w_1) p_0(w_2)}$$

$$= N_1 \sum_{j=1}^{s} p_{1j} \log \frac{p_{1j}}{p_{0j}} + N_2 \sum_{j=1}^{s} p_{2j} \log \frac{p_{2j}}{p_{0j}}. \qquad (4.3\text{-}14)$$

Similarly,

$$J(H_1, H_2) = N_1 \sum_{j=1}^{s} (p_{1j} - p_{0j}) \log \frac{p_{1j}}{p_{0j}}$$

$$+ N_2 \sum_{j=1}^{s} (p_{2j} - p_{0j}) \log \frac{p_{2j}}{p_{0j}} \qquad (4.3\text{-}15)$$

where $p_k(w_i)$ is the multinomial density with parameter vector (p_k) and frequency vector (w_i), $k = 0, 1, 2$; $i = 1, 2$.

A fundamental principle in information theory that forms the basis for most inferential statements is that of minimum-discrimination information. For a given fixed $p_0(w_i)$ the multinomial distribution, namely, $p^*(w_i)$, called the *conjugate distribution*, which minimizes

$$\sum_{(w_i)} p^*(w_i) \log \frac{p^*(w_i)}{p_0(w_i)} \tag{4.3-16}$$

such that $E^*(W_{ij}) = \theta^*_{ij} = N_i p^*_{ij}$, is given by

$$\frac{N_i!}{w_{i1}! \cdots w_{i2}!} (p_{i1}^*)^{w_{i1}} (p^*_{i2})^{w_{i2}} \cdots (p^*_{is})^{w_{is}} \tag{4.3-17}$$

where

$$p_{i1}^* = \frac{p_{ij} e^{\tau_{ij}}}{p_{i1} e^{\tau_{i1}} + \cdots + p_{is} e^{\tau_{is}}} \tag{4.3-18}$$

$j = 1, 2, \ldots, s$, and where τ_{ij} are real parameters.

Kullback [1959] has shown that if in $I(H_1:H_2)$ and in $I(H_2:H_1)$ the parameters defining the conjugate distribution replace p_{1j} and p_{2j}, and in turn these are estimated by their minimum variance unbiased estimates, and further if p_{0j} is replaced by the pooled estimate $(w_{1j} + w_{2j})/(N_1 + N_2)$, then the estimate of the error component of the divergence

$$\hat{J} = N_1 \Sigma \left(\frac{w_{1j}}{N_1} - \frac{w_{1j} + w_{2j}}{N_1 + N_2} \right) \log \frac{(N_1 + N_2) w_{1j}}{N_1 (w_{1j} + w_{2j})}$$

$$+ N_2 \Sigma \left(\frac{w_{2j}}{N_2} - \frac{w_{1j} + w_{2j}}{N_1 + N_2} \right) \log \frac{(N_1 + N_2) w_{2j}}{N_2 (w_{1j} + w_{2j})} \tag{4.3-19}$$

is under H_2 asymptotically χ^2 with $s - 1$ degrees of freedom.

To emphasize that $\hat{J}$ is dependent on the original p dichotomous variables that generated the s-state multinomial distributions in groups G_1 and G_2 we write $\hat{J} = \hat{J}(Z_1, Z_2, \ldots, Z_p)$. For any subset $Z_{i_1}, Z_{i_2}, \ldots, Z_{i_r}, r < p$ of these variables define the estimated error component of the divergence based on only these variables by $\hat{J}(Z_{i_1}, Z_{i_2}, \ldots, Z_{i_r})$; that is, act as if these are the only variables that are available to measure. Suppose we now define an ordering and a renumbering of all p variables by

$$\hat{J}(Z_1) = \max_{1 \le j \le p} \hat{J}(Z_j)$$

$$\hat{J}(Z_1 Z_2) = \max_{2 \le j \le p} \hat{J}(Z_1 Z_j)$$

$$\hat{J}(Z_1 Z_2 Z_3) = \max_{3 \le j \le p} \hat{J}(Z_1 Z_2 Z_j)$$

(etc.).   (4.3-20)

In addition, define the conditional divergences

$$\hat{J}(Z_2 | Z_1 = j_1), \quad j_1 = 0, 1$$

$$\hat{J}(Z_3 | Z_2 = j_2, Z_1 = j_1), \quad j_1 = 0, 1; \quad j_2 = 0, 1$$

$$\hat{J}(Z_4 | Z_3 = j_3, Z_2 = j_2, Z_1 = j_1), \quad j_1 = 0, 1; \quad j_2 = 0, 1; \quad j_3 = 0, 1$$

(etc.)   (4.3-21)

where, by

$$\hat{J}(Z_k | Z_{k-1} = j_{k-1}, Z_{k-2} = j_{k-2}, \ldots, Z_1 = j_1), \qquad (4.3\text{-}22)$$

we mean the estimated divergence computed from the conditional distribution of Z_k given Z_1 is at level j_1, Z_2 is at level $j_2,\ldots,$ and Z_{k-1} is at level j_{k-1}.

Under the hypothesis that in both groups the binomial distributions induced by Z_j are the same, $\hat{J}(Z_j)$ is asymptotically χ^2 with one degree of freedom. Similarly, under the hypothesis that in both groups the multinomial distributions induced by the pair (Z_j, Z_k), $j \ne k$, are identical, $\hat{J}(Z_j Z_k)$ is asymptotically χ^2 with three degrees of freedom, and so on. It does not follow, however, that $\hat{J}(Z_1) = \max \hat{J}(Z_j)$ and $\hat{J}(Z_1 Z_2) = \max_{j \ne 1} \hat{J}(Z_1 Z_j)$ are also asymptotically χ^2 with one and three degrees of freedom, respectively. However, in describing the selection procedure we use the critical values for χ^2 when making decisions as to where to stop given the magnitude of the various ordered $\hat{J}$ values. This approach can lead to more variables or levels of variables being included than perhaps would be necessary on the basis of the strict distributional properties of the max $\hat{J}$ statistics. Note that this is similar in spirit to the idea in the forward selection procedure in stepwise regression where the variable considered for inclusion at any given step is the one yielding the largest single degree of freedom F-ratio among those eligible for inclusion. To reduce the chance of including variables or levels of variables that may not be necessary given the true distributions of the max $\hat{J}$ values, the researcher may wish to decrease the specified α-level or not include cases where marginal significance is obtained.

After all variables are renumbered, the selection procedure proceeds by observing Z_1 and computing $\hat{J}(Z_1)$. If on the basis of the sample sizes N_1 and N_2, $\hat{J}(Z_1)$ is significant, observe Z_2 and compute $\hat{J}(Z_1 Z_2)$. If this is significant, compute the two estimated conditional divergences $\hat{J}(Z_2|Z_1=0)$ and $\hat{J}(Z_2|Z_1=1)$. Both statistics are then compared to a specified χ^2 value with one degree of freedom. If, for example, $\hat{J}(Z_2|Z_1=0)$ is significant but not $\hat{J}(Z_2|Z_1=1)$, then observe Z_2 only if $Z_1=0$.

The procedure continues by computing $\hat{J}(Z_1 Z_2 Z_3)$, and if significant when compared to χ^2 with seven degrees of freedom computing the four conditional divergences $\hat{J}(Z_3|Z_1=j_1, Z_2=j_2), j_1=0,1; j_2=0,1$. Variable Z_3 is then only observed with those levels of Z_1 and Z_2 for which $\hat{J}(Z_3|Z_1=j_1, Z_2=j_2)$ is significant. Variables cease to enter the system at the rth $(r>1)$ stage if $\hat{J}(Z_1, Z_2, \ldots, Z_r)$ is nonsignificant. Further, the conditional divergence computed up through stage $r-1$ indicates whether a new variable is worthy of entry based on the levels of those previously observed. Although we use the term *stepwise* for this procedure, it is really a forward procedure as mentioned above in that there is no provision made for deletion of a variable once it has been entered.

The stepwise procedure suggested here is similar in a sense to Hills's restricted and unrestricted stepwise procedure in that levels of good discriminatory variables are split by new variables to see if their inclusion in a given classification rule improves separation of the groups. The principal difference, however, is that whereas Hills uses the divergence statistic only to rank variables for entry, it is proposed here to utilize distributional properties of a minimum discrimination divergence statistic to test whether the inclusion of a new variable yields a result consistent with the hypothesis that multinomial distributions so induced in G_1 and G_2 are identical. The logic for not including a variable is that it is reasonable to assume that discriminatory power will not be increased sufficiently if the divergence is not large enough to reject the hypothesis that at a given stage the two groups have identical multinomial distributions.

4.4   AN ILLUSTRATIVE EXAMPLE

To better understand the variable-selection schemes proposed for finding good subsets of variables, three of the four selection procedures discussed in this chapter are illustrated on a common set of data. The three variable-selection procedures illustrated are the Goldstein and Rabinowitz [1975] procedure and the two sequential procedures due to Lachin [1973] and Goldstein and Dillon [1977]. As we indicated, the Raiffa [1961] sequential procedure is not demonstrated with data since at present it is not computer-operational.

4.4.1 The Data

Data collected in a 1975 study designed to examine the consumption behavior for a major household service product illustrate the respective variable-selection procedures. Respondents were selected from three geographically dispersed areas in which the sponsoring company offered its services. The grouping variable under consideration is the extent of customer usage, where usage is in terms of actual dollar expenditures for a fixed time period. Of a total of 464 respondents, 234 were identified as "heavy users," and the remaining 230 were identified as "light users." The distinction between heavy and light users conformed to the firm's general

TABLE 4.4-1

VARIABLE CODINGS AND DESCRIPTIONS

	DESCRIPTIONS	
VARIABLE	CODING 1	CODING 0
Home ownership (x_1)	Own	Rent
Number of rooms in home (x_2)	At least five rooms	Less than five rooms
Length of residence (x_3)	At least five yr	Less than five yr
Location of previous home (x_4)	Outside of county, state, or country (U.S.A.)	Within the same town or county
Marital status (x_5)	Married	Single, widowed, or divorced
Head of household's occupation (x_6)	Professional or manager	Sales, craftsman, clerical worker (or below)
Head of household's education (x_7)	At least some college	No college
Family income (x_8)	At least \$11,000/yr	Less than \$11,000/yr
Stage in family life cycle (x_9)	Age 55 yr or older, employed or unemployed	Less than 55 yr old

experiences based on extensive market segmentation research. Nine socio-economic and demographic variables were selected to form the basis for classification.

In our discussion of the common approach to the analysis of qualitative data in Section 1.3.1, we presented Table 1.3-1, which described the nine predictor variables collected in this 1975 study. Referring to the table we see that as originally defined the nine categoric variables ranged from two levels for X_1—home ownership, to 11 levels for X_8—family income. Obviously, if all levels are maintained an immense number of states relative to the sample size would result, making estimation of the state probabilities highly unreliable. To deal with this problem it was decided to dichotomize all variables, thereby reducing the number of response patterns to 512 by determining a critical cutoff point for each variable such that the difference between the relative cumulative frequencies across the two groups is a maximum. Table 4.4-1 details the coding scheme that resulted.

4.4.2 Application of the Procedures

Method 1—The Procedure Based on Differences in Discriminant Scores (Goldstein and Rabinowitz [1975]). Recall that the Goldstein–Rabinowitz selection procedure is based on a pseudo distance d that in some sense is a measure of distance between discriminant scores $g_1(x)$ and $g_2(x)$. In terms of the discriminant scores, two conditions were proposed by these authors for the selection of variates problem. Condition (4.2-3) selects that subset of variables that maximizes the estimated scaled value of d, and condition (4.2-4) selects that subset of variables that maximizes the scaled average value of $|(N_1(x(i)))^{\frac{1}{2}} - (N_2(x(i)))^{\frac{1}{2}}|$.

Table 4.4-2 presents summary information concerning the estimated maximum scaled minimum and average values of $|\sqrt{g_1} - \sqrt{g_2}|$ for subsets of size n. Use of conditions (4.2-3) and (4.2-4) requires that we maximize over these nine values. In terms of conditions (4.2-3) maximizing over these values indicates that the optimal subset contains all variables, whereas according to condition (4.2-4) the optimal subset in this sense includes seven variables, namely, $(x_2, x_3, x_5, x_6, x_7, x_8, x_9)$.

To determine how well we do using the two respective conditions, we classified all 464 respondents on the basis of the full multinomial rule. Using the entire set of variables indicated by condition (4.2-3) results in 117 misses or an apparent error of 25.2%. On the other hand, using the seven predictors indicated by condition (4.2-4) results in 143 misses or an apparent error of 30.8%. In terms of error rates, then, condition (4.2-3) does better than condition (4.2-4); however, this is not surprising since in general the probability of misclassification decreases as the number of

TABLE 4.4-2

Estimation of the Maximum Scaled Minimum and Average
Values of $\left|\sqrt{g_1} - \sqrt{g_2}\right|$ for Subsets of Size n

SUBSET SIZE n	$x(i)$ SELECTED	MAXIMUM SCALED MINIMUM $\left|\sqrt{N_1(x(i))} -\sqrt{N_2(x(i))}\right|$	$x(i)$ SELECTED	MAXIMUM SCALED AVERAGE $\left|\sqrt{N_1(x(i))} -\sqrt{N_2(x(i))}\right|$
9	$(x_1,x_2,x_3,x_4,x_5,x_6,x_7,x_8,x_9)$	0.19626	$(x_1,x_2,x_3,x_4,x_5,x_6,x_7,x_8,x_9)$	0.25308
8	$(x_1,x_2,x_3,x_4,x_5,x_7,x_8,x_9)$	0.18947	$(x_1,x_2,x_4,x_5,x_6,x_7,x_8,x_9)$	0.28890
	$(x_1,x_3,x_4,x_5,x_6,x_7,x_8,x_9)$	0.18947		
7	$(x_1,x_2,x_3,x_4,x_6,x_8,x_9)$	0.15892	$(x_2,x_3,x_5,x_6,x_7,x_8,x_9)$	0.29846
6	$(x_3,x_4,x_5,x_6,x_8,x_9)$	0.11237	$(x_2,x_4,x_5,x_6,x_8,x_9)$	0.28099
5	(x_1,x_2,x_4,x_5,x_6)	0.12683	(x_2,x_5,x_6,x_8,x_9)	0.25620
	(x_3,x_4,x_5,x_6,x_7)	0.12683		
	(x_1,x_3,x_4,x_7,x_9)	0.12683		
	(x_4,x_5,x_7,x_8,x_9)	0.12683		
4	(x_3,x_4,x_5,x_6)	0.12683	(x_4,x_5,x_8,x_9)	0.22370
3	(x_3,x_5,x_6)	0.08909	(x_4,x_5,x_8)	0.19444
2	(x_5,x_8)	0.13012	(x_5,x_8)	0.18419
1	(x_8)	0.14796	(x_8)	0.18081

82

variables included in the predictor set is increased. Furthermore, at least from a practical perspective, the minimization process is of no value since the entire set of predictors is included, and hence the problems that probably motivated us to consider variable-selection schemes remain present. We have more to say about the relative advantages and disadvantages of the d measure in the next section.

Method 2—Lachin's [1973] Procedure. Lachin's stepwise selection procedure chooses variables on the basis of the conditional independence of X_0, having two levels that define the characteristics of the two groups of interest, and an additional predictor X_{t+1} given the levels of the remaining variables $(X_1, X_2, \ldots, X_t)$. In effect, this method measures the change in the degree of association between X_0 and the remaining variables when X_{t+1} is added to the predictor set. Recall that the stepwise procedure at any given stage is dependent on the magnitude of the statistic $\chi^2_{(3)}$ and requires that the minimum acceptable probability for the addition of a variable (MPA) and the minimum acceptable probability level for the deletion of a variable already selected (MPD) be specified.

Table 4.4-3 summarizes the results for Lachin's procedure at the last step using an MPA and MPD of 0.10. The process starts by selecting X_8—family income and then terminates after selecting X_6—head-of-household's occupation. With only two variables included in the predictor set 174 misses occur (using the full multinomial rule), or an apparent error of 37.5%. It is interesting to note that X_6—head of household's occupation does not increase discrimination in that use of X_8—family income would

TABLE 4.4-3

RESULTS FOR LACHIN'S PROCEDURE AT THE LAST STEP (MPA AND MPD = 0.10)

VARIABLES SELECTED	VARIABLES NOT SELECTED	χ_i^2	DF	$\Delta \chi_i^2$	ΔDF	$P(\Delta \chi_i^2)$
X_8		19.85	1	20.51	2	0.35 E-04
X_6		33.48	1	6.89	2	0.32 E-01
	X_1	40.65	7	0.28	4	0.99
	X_2	42.43	7	2.07	4	0.72
	X_3	44.15	7	3.79	4	0.44
	X_4	46.89	7	6.52	4	0.16
	X_5	44.70	7	4.34	4	0.36
	X_7	42.24	7	1.88	4	0.76
	X_9	47.13	7	6.76	4	0.15

TABLE 4.4-4

RESULTS FOR LACHIN'S PROCEDURE AT THE LAST STEP (MPA AND MPD = 0.20)

VARIABLES SELECTED	VARIABLES NOT SELECTED	χ_i^2	DF	$\Delta\chi_i^2$	ΔDF	$P(\Delta\chi_i^2)$
X_8		32.55	7	30.04	8	0.21 E-03
X_6		43.31	7	19.28	8	0.13 E-01
X_9		44.70	7	17.89	8	0.22 E-01
X_5		47.13	7	15.46	8	0.51 E-01
	X_1	72.30	31	9.71	16	0.88
	X_2	76.20	31	13.62	16	0.63
	X_3	73.90	31	11.31	16	0.79
	X_4	78.07	29	15.48	14	0.35
	X_7	70.30	31	7.71	16	0.96

yield an identical error rate of 37.5%. From a practical perspective, the process appears too restrictive in that only two variables are selected and, therefore, a practitioner might well question whether relevant information is being discarded. Obviously, this problem can be handled by setting a less conservative value for the MPA. For example, Table 4.4-4 shows the results of applying Lachin's procedure using an MPA and MPD of 0.20. The procedure terminated at the fifth step with four predictors having been selected: (a) X_8—family income, (b) X_6—head of household's occupation, (c) X_9—family life cycle, and (d) X_5—marital status. In this application it is interesting to note that X_5 was selected although it did not show a significant association with heavy or light users initially. As expected, use of five variables yielded a slightly better apparent error—approximately 35%.

Method 3—The Procedure Based on a Kullback Divergence Statistic (Goldstein and Dillon [1977]). Recall from Section 4.3.3 that the Goldstein–Dillon stepwise selection procedure uses the distributional properties of a Kullback minimum discrimination divergence statistic not only to provide stopping rules for the inclusion of new variables but also to determine the contribution of new variables given the levels of those variables already included. Hence in this sense it represents a substantial improvement over other forward selection procedures.

Table 4.4-5 shows those variables that enter using the group-divergence stopping rule. We see from the table that the process terminates at the fifth step with variables X_8—family income, X_6—head-of-household's occupation, X_4—location of previous home, X_9—family life cycle, and X_5—marital status having been selected. The apparent errors given in Table 4.4-5 were again computed using the full multinomial rule. When all nine variables are used, the apparent error evaluates to 25.2% as compared to 32.7% when only the five selected variables are used. It should be noted that if α were set to 0.10 the procedure would have included X_2—number of rooms in home and then terminate.

As we indicated earlier, this procedure affords the practitioner the ability to determine the contribution of new variables given the levels of those variables already in the system. Table 4.4-6 presents the conditional divergences, and at each step a decision is made as to the importance of the new variable given the levels of those variables previously included. For example, after X_6 is entered the two conditional divergences $\hat{J}(X_6|X_8=0)$ and $\hat{J}(X_6|X_8=1)$ are computed and since both statistics are significant ($\alpha=0.05$), we conclude that head-of-household's occupation is important in the presence of family income at both its levels. At the third step variable X_4—location of previous home is selected, but it is interesting that Table 4.4-6 indicates that this variable does not contribute enough divergence in the presence of $X_6=1$ and $X_8=0$ to warrant its inclusion for discrimination. In other words, no significant amount of discriminatory information is obtained from examining responses to the location of previous home question if we know that the head-of-household's occupation is either professional or manager and if family income is less than $11,000 a year. It should be noted that after the fourth step the conditional divergences of X_9 given $X_4=(0,1)$, $X_6=1$, and $X_8=0$ are not shown since at the previous step the conditional divergences for patterns $(X_4=0,X_6=1,X_8=0)$ and $(X_4=1,X_6=1,X_8=0)$ were nonsignificant. Similarly, after

TABLE 4.4-5

VARIABLES ENTERING USING GROUP-DIVERGENCE STOPPING RULE

VARIABLE SELECTED	$\hat{J}$	DF	CRITICAL VALUE (0.05)	APPARENT ERROR
X_8	34.92	1	3.84	0.375
X_6X_8	43.23	3	7.81	0.375
$X_4X_6X_8$	51.15	7	14.07	0.373
$X_9X_4X_6X_8$	61.83	15	25.00	0.347
$X_5X_9X_4X_6X_8$	86.00	31	44.97	0.327
Stop				

TABLE 4.4-6

LEVELS OF VARIABLES ENTERING USING CONDITIONAL-DIVERGENCE
STOPPING RULE

VARIABLE SELECTED GIVEN PREVIOUS LEVELS	$\hat{J}$	DF	CRITICAL VALUE (0.05)
$X_6 \| X_8 = 0$	17.80	1	3.84
$X_6 \| X_8 = 1$	4.30	1	3.84
$X_4 \| X_6 = 0, X_8 = 0$	10.80	1	3.84
$X_4 \| X_6 = 0, X_8 = 1$	13.50	1	3.84
$X_4 \| X_6 = 1, X_8 = 0$	1.17	1	3.84
$X_4 \| X_6 = 1, X_8 = 1$	5.10	1	3.84
$X_9 \| X_4 = 0, X_6 = 0, X_8 = 0$	14.50	1	3.84
$X_9 \| X_4 = 0, X_6 = 0, X_8 = 1$	16.00	1	3.84
$X_9 \| X_4 = 0, X_6 = 1, X_8 = 1$	1.50	1	3.84
$X_9 \| X_4 = 1, X_6 = 0, X_8 = 0$	48.50	1	3.84
$X_9 \| X_4 = 1, X_6 = 0, X_8 = 1$	11.67	1	3.84
$X_9 \| X_4 = 1, X_6 = 1, X_8 = 1$	23.70	1	3.84
$X_5 \| X_4 = 0, X_6 = 0, X_8 = 0, X_9 = 0$	85.37	1	3.84
$X_5 \| X_4 = 0, X_6 = 0, X_8 = 0, X_9 = 1$	99.57	1	3.84
$X_5 \| X_4 = 0, X_6 = 0, X_8 = 1, X_9 = 1$	3.28	1	3.84
$X_5 \| X_4 = 1, X_6 = 0, X_8 = 1, X_9 = 1$	0.03	1	3.84

step 5 several divergences computed from the conditional distribution of X_5 for given patterns of X_4, X_6, X_8, and X_9 were nonsignificant and hence are also omitted from Table 4.4-6. In addition, there are five patterns that dropped out because of state sparseness. To summarize, we see that the stepwise selection procedure reduces the number of variables from nine to five and, more dramatically, we need now consider only those 24 states that provide significant divergence instead of the original 512. The patterns that provided significant divergence along with their description are shown in Table 4.4-7 (pp. 88–89).

4.4.3 Some Additional Considerations

Each of the variable selection procedures resulted in a different set of variates being selected. However, in the case of the two sequential procedures some degree of consistency was evident in that both selected X_8—family income at the first step followed by X_6—head-of-household's occupation. The minimization process given by condition (4.2-3) appears least useful in that all variables are selected. It is recommended that extreme care be exercised when using this procedure since it is possible to obtain misleading results especially when the number of sample observations used in the estimation process is limited. That is, in applying condition (4.2-3) we are in the uncomfortable position of relying solely on an estimate that can be highly unstable and, therefore, reducing both frequency distributions for a given set of variables by considering only the scaled minimum of $\left| (N_1(x(i)))^{\frac{1}{2}} - (N_2(x(i)))^{\frac{1}{2}} \right|$ discards much information that becomes more acute as the sample sizes decrease. Since the average of $\left| (N_1(x(i)))^{\frac{1}{2}} - N_2(x(i))^{\frac{1}{2}} \right|$ uses more information from the frequency distributions involved, it appears less subject to the irregularities present in a particular data set and in this sense is preferable to the minimization process.

Although Lachin's procedure included the same variables as the Goldstein–Dillon sequential procedure, it terminated after only two steps; hence it appears far too conservative. Furthermore, application of this procedure to the data described in Section 2.4 proved similar in that it again resulted in only a few variables being selected. From these results it would appear that more liberal inclusion levels should be considered. This recommendation is in a sense similar to the results reported by Bendel and Afifi [1977], who found that for a variety of stopping rules in forward stepwise regression α levels between 0.15 and 0.25 were superior.

The ability to reduce the dimensionality of the problem, where the selected subset of variables contains more than one or two variates, is particularly important in many areas of applied research. In this sense the Goldstein–Dillon stepwise procedure appears most useful in that it has the ability to identify specific response categories that warrant consideration. Hence by allowing the practitioner the opportunity not only to select good subsets of variables but also to identify those levels that need measurement, the practitioner is in a position to significantly reduce the cost and effort of collecting measurements in subsequent studies.

4.5   AFTERTHOUGHT

Care and good judgment on the part of the statistician is an essential prerequisite in the intelligent application of any forward or stepwise procedure, whether a "best" regression equation or discriminant rule is the objective. As we stated in the beginning of this chapter, pockets of sparseness in a data set frequently are the motivating force in considering variable-selection schemes. However, the same problem of sparseness that led us to consider such procedures is like an inescapable fugue always reappearing to remind us of its presence.

TABLE 4.4-7

DESCRIPTION OF RESPONSE PATTERNS PROVIDING SIGNIFICANCE DIVERGENCES

$X_8 X_6 X_4 X_9 X_5$	FAMILY INCOME	HEAD OF HOUSE-HOLD'S OCCUPATION	LOCATION OF PREVIOUS HOME	FAMILY LIFE CYCLE	MARITAL STATUS
0 0	<$11,000	Sales, craftsman, Clerical (or below)			
1 0	>$11,000	Sales, craftsman, clerical (or below)			
0 1	<$11,000	Professional or manager			
1 1	>$11,000	Professional or manager			
0 0 0	<$11,000	Sales, craftsman, clerical (or below)	Within same town or county		
1 0 0	>$11,000	Sales, craftsman, clerical (or below)	Within same town or county		
1 1 0	>$11,000	Professional or manager	Within same town or county		
0 0 1	<$11,000	Sales, craftsman, clerical (or below)	Outside of county, state, or country (U.S.A.)		
1 0 1	>$11,000	Sales, craftsman, clerical (or below)	Outside of county, state, or country (U.S.A.)		
1 1 1	>$11,000	Professional or manager	Outside of county, state, or country (U.S.A.)		
0 0 0 0	<$11,000	Sales, craftsman, clerical (or below)	Within same town or county	Less than 55 years old	
1 0 0 0	>$11,000	Sales, craftsman, clerical (or below)	Within same town or county	Less than 55 years old	

Code	Income	Occupation	Location	Age/Status	Marital Status
0 0 1 0	<$11,000	Sales, craftsman, clerical (or below)	Outside of county, state, or country (U.S.A.)	Less than 55 years old	
1 0 1 0	>$11,000	Sales, craftsman, clerical (or below)	Outside of county, state, or country (U.S.A.)	Less than 55 years old	
1 1 1 0	>$11,000	Professional or manager	Outside of county, state, or country (U.S.A.)	Less than 55 years old	
0 0 0 1	<$11,000	Sales, craftsman, clerical (or below)	Within same town or county	At least 55 years old, employed or unemployed	
1 0 0 1	>$11,000	Sales, craftsman, clerical (or below)	Within same town or county	At least 55 years old, employed or unemployed	
0 0 1 1	<$11,000	Sales, craftsman, clerical (or below)	Outside of county, state, or country (U.S.A.)	At least 55 years old, employed or unemployed	
1 0 1 1	>$11,000	Sales, craftsman, clerical (or below)	Outside of county, state, or country (U.S.A.)	At least 55 years old, employed or unemployed	
1 1 1 1	>$11,000	Professional or manager	Outside of county, state, or country (U.S.A.)	At least 55 years old, employed or unemployed	
0 0 0 0 0	<$11,000	Sales, craftsman clerical (or below)	Within same town or county	Less than 55 years old	Single, widowed, or divorced
0 0 0 1 0	<$11,000	Sales, craftsman, clerical (or below)	Within same town or county	At least 55 years employed or unemployed	Single, widowed, or divorced
0 0 0 0 1	<$11,000	Sales, craftsman, clerical (or below)	Within same town or county	Less than 55 years old	Married
0 0 0 1 1	<$11,000	Sales, craftsman, clerical (or below)	Within same town or county	At least 55 years old, employed or unemployed	Married

The Goldstein–Dillon procedure may need to be amended (unless there are many observations available) after a number of variables have been entered. Problems will arise in generating stable divergence estimates derived from considering high-order conditional distributions if state frequencies are very small. Similarly, Lachin's [1973] stepwise method is also subject to difficulties of consistency and reliability at points where state frequencies are so low as to render χ^2 statistics of questionable validity. It is in such situations that the formal probabilistic stopping rules have to be interpreted softly, and not as stringent tolerances forcing us one way or the other. In addition, it is recommended that a validation sample be used to get a better estimate of the worthiness of the variables selected by a particular procedure. Moreover, it is obvious that the performance of a given subset of variables, say, in terms of error rates, will in general be dependent on the choice of an appropriate classification rule. Therefore, we further recommend that alternative model specifications such as those discussed in Chapter 2 be considered along with the decision as to which variable-selection procedure is most appropriate.

CHAPTER 5

Special Topics

5.1 INTRODUCTION

There are numerous topics related to the discrete discriminant problem that deserve special attention but not separate chapters. This chapter, therefore, is devoted to a discussion of several rather unconnected topics that merit consideration. The topics or special problem areas to be discussed include (a) methods for assessing the relative effectiveness within a given data set of competing classification procedures, (b) a classification method for handling data sets composed of both discrete and continuous variables, and (c) an overview of several studies that have utilized a Monte Carlo sampling framework to assess the relative performance of a number of classification procedures on a wide variety of population structures.

5.2 MIXTURE OF VARIABLES

All models discussed in Chapter 2 specifically address the issue of classification with discrete vector observations. In particular, when dealing with multivariate dichotomous responses we are in a sense considering a class of problems most removed from data assumed generated by multivariate normal distributions. However, in many situations researchers must deal with data having both discrete and continuous variables and not strictly with one or the other. Classification procedures that consider the problem of joining both continuous and discrete techniques to effect a rule for classification have only recently received attention in the literature; however, it is an area of investigation worthy of much additional study.

Krzanowski [1975, 1976, 1977] recently considered the following model. Let X be a q-dimensional binary variable generating $k = 2^q$ states, and suppose Y is a p-dimensional normal random variable having mean $\mu_i^{(m)}$ in state m and population Π_i ($m = 1, 2, \ldots, k$; $i = 1, 2$) and common variance–covariance matrix Σ in all states of both populations. Note that the model states that conditional on X, the variance–covariance matrix in Π_1 and Π_2 are equal but the unconditional matrices are not.

If we let $\mathbf{W}=(\mathbf{X},\mathbf{Y})$ and denote the probability density of $\mathbf{W}$ in Π_i by $p_i(\mathbf{w})=p_i(\mathbf{y}|\mathbf{x})p_i(\mathbf{x})$, then assuming equal prior probabilities the optimal rule as defined in Chapter 2 is to allocate to Π_1 if

$$\frac{p_1(\mathbf{y}|\mathbf{x})}{p_2(\mathbf{y}|\mathbf{x})} \geqslant \frac{p_2(\mathbf{x})}{p_1(\mathbf{x})} \tag{5.2-1}$$

and otherwise to Π_2. Assuming that the fixed realization $\mathbf{x}$ is in state m, and denoting $p_i(\mathbf{x})$ by p_{im}, it follows that the partition given in (5.2-1) reduces to

$$\left(\boldsymbol{\mu}_1^{(m)}-\boldsymbol{\mu}_2^{(m)}\right)'\boldsymbol{\Sigma}^{-1}\left\{\mathbf{y}-\frac{1}{2}\left(\boldsymbol{\mu}_1^{(m)}+\boldsymbol{\mu}_2^{(m)}\right)\right\} \geqslant \ln\left(p_{2m}/p_{1m}\right). \tag{5.2-2}$$

Note that the optimum rule reduces to k different linear discriminant functions, one for each of the $k=2^q$ multinomial states. When $k=2$ (the case of one binary variable), (5.3-2) reduces to the rule considered by Chang and Afifi [1972].

Assuming that all parameters are known, the two probabilities of misclassification $P(2|1)$ and $P(1|2)$ where $P(i|j)$ is the probability of assignment to Π_i when the observation comes from Π_j are given by

$$P(2|1)=\sum_{m=1}^{k} p_{1m}\Phi\left\{\frac{\ln(p_{2m}/p_{1m})-1/2D_m^2}{D_m}\right\}$$

$$P(1|2)=\sum_{m=1}^{k} p_{2m}\Phi\left\{\frac{\ln(p_{1m}/p_{2m})-1/2D_m^2}{D_m}\right\} \tag{5.2-3}$$

where D_m^2 is the Mahalanobis distance $(\boldsymbol{\mu}_1^{(m)}-\boldsymbol{\mu}_2^{(m)})'\boldsymbol{\Sigma}^{-1}(\boldsymbol{\mu}_1^{(m)}-\boldsymbol{\mu}_2^{(m)})$ and where $\Phi(z)$ is the standard normal CDF at z. If $k=1$ and $p_{1k}=p_{2k}$, then the probabilities of misclassification reduce to the usual expressions $P(2|1)=P(1|2)=\Phi(-1/2D)$, where D is the one Mahalanobis distance that remains since there is now only one discriminant function.

Estimation of the optimal rule can proceed in the usual way with initial samples of size n_i from $\Pi_i, i=1, 2$. Let n_{im} be the number of observations out of n_i that belong to state m and denote by $\mathbf{y}_{ji}^{(m)}$ the vector of continuous measurements associated with the jth observation in state m from Π_i. Then, if

$$\bar{\mathbf{y}}_i^{(m)}=\frac{1}{n_{im}}\sum_{j=1}^{n_{im}} \mathbf{y}_{ji}^{(m)}, \tag{5.2-4}$$

the maximum likelihood estimates of p_{im}, $\mu_i^{(m)}$ and Σ are given by

$$\hat{p}_{im} = \frac{n_{im}}{n_i}, \quad \hat{\mu}_i^{(m)} = \bar{y}_i^{(m)}$$

$$\hat{\Sigma} = \frac{1}{(n_1 + n_2 - 2k)} \sum_{i=1}^{2} \sum_{m=1}^{k} \sum_{j=1}^{n_{im}} \left(y_{ji}^{(m)} - \bar{y}_i^{(m)}\right)\left(y_{ji}^{(m)} - \bar{y}_i^{(m)}\right) \qquad (5.2\text{-}5)$$

$(i = 1,2; \ m = 1,2,\ldots,k)$. The estimate $\hat{\Sigma}$ is just the expression for the pooled-covariance matrix assuming as we are that the $2k$ matrices are equal.

Although the estimates given in (5.2-5) are satisfactory when sufficient numbers of observations in each state are available, what usually happens, as we have repeatedly discussed, is that many cells either are empty or have too few observations to generate reasonable estimates. Krzanowski suggests fitting a suitable loglinear model to estimate state probabilities, to wit; it is assumed that the observed frequency n_{im} in state m of the sample from Π_i is a realization of a variate having nonzero mean v_{im} expressible as

$$\ln v_{im} = \sum_{j=1}^{s} a_{imj} \theta_j \qquad (5.2\text{-}6)$$

for some set of unknown constants $\theta_1, \theta_2, \ldots, \theta_s$ and known coefficients $a_{im1}, a_{im2}, \ldots, a_{ims}$. The θ constants are interpreted to represent main effects of the binary variables and interactions between them of all orders up to q. Usually, however, only terms up through first-order interactions are maintained prior to utilizing an iterative proportional fitting algorithm. However, if the states are very sparse, then one may opt to use only main effects. In any case if n_{im} represents the nonzero estimate obtained, then the state estimate $\hat{p}_{im} = (n_{im}/n_i)$ is used.

State sparseness also affects the quality or availability of estimates for $\mu_i^{(m)}$ and Σ. For this part of the problem Krzanowski utilized a linear additive model, the parameters of which are again interpreted in terms of main effects and interaction terms. In particular, a model of the form

$$\mu_i = v_i + \sum_{j=1}^{q} \alpha_{j,i} x_j + \sum_{j<k} \sum \beta_{jk,i} x_j x_k$$

$$+ \cdots + \delta_{1,2\cdots q} x_1 x_2 \cdots x_q \qquad (5.2\text{-}7)$$

is suggested. The conditional mean vector in state m, $\mu_i^{(m)}$ is obtained by inserting the values of the binary variables in state m into the right-hand side of (5.2-7).

One usually would only fit a first-order interaction model, thereby only maintaining the parameters ν_i, $\alpha_{j,i}$, and $\beta_{jk,i}$. In this case if we write

$$\boldsymbol{\mu}' = (1, x_1, x_2, \ldots, x_q, x_{12}, x_{13}, \ldots, x_{q-1q})$$

and

$$\mathbf{R}_i = (\nu_i, \alpha_{1,i}, \alpha_{2,i}, \ldots, \alpha_{q,i}, \beta_{12,i}, \ldots, \beta_{q-1q,i}) \qquad (5.2\text{-}8)$$

then the conditional density of Y given X is normal with mean $\mathbf{R}_i\boldsymbol{\mu}$ and variance $\hat{\boldsymbol{\Sigma}}$. Maximum likelihood estimates $\hat{\mathbf{R}}_i$ and $\hat{\boldsymbol{\Sigma}}$ are given by

$$\hat{\mathbf{R}}_i = \mathbf{C}_i \mathbf{A}_i^{-1}$$

$$(n_1 + n_2)\hat{\boldsymbol{\Sigma}} = \sum_{i=1}^{2} \sum_{j=1}^{n_i} \left\{ \mathbf{y}_{ji} \mathbf{y}'_{ji} - \hat{\mathbf{R}}_i \mathbf{A}_i \hat{\mathbf{R}}'_i \right\} \qquad (5.2\text{-}9)$$

where

$$\mathbf{C}_i = \Sigma_{j=1}^{n_i} \mathbf{y}_{ji} \boldsymbol{\mu}'_{ji}$$

and

$$\mathbf{A}_i = \Sigma_{j=1}^{n_i} \boldsymbol{\mu}_{ji} \boldsymbol{\mu}'_{ji},$$

provided that $\mathbf{U}_i = (\boldsymbol{\mu}_{1i}, \ldots, \boldsymbol{\mu}_{ni})$ is of rank $1 + q + 1/2q(q-1)$, and that $n_i \geqslant 1 + p + 1/2q(q+1)$.

Krzanowski also considers the estimation of error rates using Lachenbruch's "leaving-our method" and demonstrates the feasibility of the calculations. In addition, he treats through error rates the ramification of considering the binary variables as if they were continuous and proceeding with LDF as the rule for classification. Although his discussion is extensive and instructive, there also are some immediate extensions to more general settings. One can, for example, immediately apply the rule given in (5.2-2) to the case where the discrete components of the W vector can assume more than a dichotomy. In general, if $X = (X_1, X_2, \ldots, X_p)$ is such that X_j has s_j levels, then the number of states in each group is given by $\Pi_{j=1}^{q} s_j$ which then becomes the number of linear discriminant functions needed to cover the contingency for classifying any future observations. However, the problems that led to considering a loglinear model for estimating state probabilities now become more extreme since pockets of sparseness will be amplified by the increase in the number of states. In theory, however, the rule for accommodating such sturctures is a straightforward extension of (5.2-2).

The location model assumes that conditional on X, Y is multivariate normal with mean vector $\mu_i^{(m)}$ in state m under Π_i and covariance matrix Σ in all states under both $\Pi_i, i = 1, 2$. If we allow for the case where conditional on X, Y has covariance matrix under Π_i given by Σ_i but the same within all states, then the rule becomes to assign to Π_1 if

$$\frac{1}{2}\left\{ \ln \frac{|\Sigma_1|}{|\Sigma_2|} - (y - \mu_1^{(m)})'\Sigma_1^{-1}(y - \mu_1^{(m)}) \right.$$

$$\left. + (y - \mu_2^{(m)})'\Sigma_2^{-1}(y - \mu_2^{(m)}) \right\} \geqslant \ln(p_{2m}/p_{1m}). \quad (5.2\text{-}10)$$

Notice that the rule is the familiar quadratic discriminant function. Estimation, if sufficient data are available, can proceed as in (5.2-4) and (5.2-5), with the exception that we no longer use the pooled estimate in (5.2-5) but estimate the covariance matrices separately by

$$\hat{\Sigma}_i = \frac{1}{(n_i - 1)} \sum_{m=1}^{k} \sum_{j=1}^{n_{im}} (y_{ji}^{(m)} - \bar{y}_i^{(m)})(y_{ji}^{(m)} - \bar{y}_i^{(m)})'. \quad (5.2\text{-}11)$$

However, for most data sets we probably would be more inclined to fit a linear additive model as in (5.2-7) for the parameters $\mu_1^{(m)}$, $\mu_2^{(m)}$, Σ_1, and Σ_2.

Perhaps the most general extension of the location model occurs when the assumption of normality for the conditional density of Y given X is relaxed. If we assume only that conditional on X, Y has continuous components that generate some joint distribution not necessarily the same in each state and different under Π_1 and Π_2, then we may utilize techniques in nonparametric density estimation in deriving a classification rule. Because we are providing for the distributions to be distinct between states within a particular population Π_i, the procedure will probably only be viable if a few binary variables are measured and "reasonably" healthy frequencies are available in each state.

Cacoullous [1966] extended ideas by Rosenblatt [1956] and Parzen [1962] to the multivariate case in deriving a family of density estimators. In general, if $Y = (Y_1, Y_2, \ldots, Y_p)$ is a p-component continuous vector, then the density estimate at $y = (y_1, y_2, \ldots, y_p)$ considered by Cacoullous is given by

$$f_n(y) = \frac{1}{n\Pi_{i=1}^{p} h_i} \sum_{j=1}^{n} K\left(\frac{y_1 - Y_{j1}}{h_1}, \frac{y_2 - Y_{j2}}{h_2}, \ldots, \frac{y_p - Y_{jp}}{h_p} \right) \quad (5.2\text{-}12)$$

where $(Y_{11}, Y_{12}, \ldots, Y_{1p}) \ldots, (Y_{n1}, Y_{n2}, \ldots, Y_{np})$ is a random sample of size n

and $h_i = h_i(n)$ are constants that approach zero as n approaches infinity. Under suitable conditions on the kernel function $K(\cdot)$ and additionally assuming $n^p \Pi_{i=1}^p h_i(n) \to \infty$, Cacoullous demonstrated that $f_n(\mathbf{y})$ is a mean-square consistent estimate of the true density $f(\mathbf{y})$. It follows, therefore, from Glick [1972] that the sample-based likelihood-ratio rule induced through utilizing $f_{n_i}(\cdot)$ in place of the true densities $f_i(\cdot)$ is a consistent procedure.

One particular kernel that is quite simple has the form

$$K\left(\frac{y_1 - Y_{j1}}{h_1}, \frac{y_2 - Y_{j2}}{h_2}, \ldots, \frac{y_p - Y_{jp}}{h_p}\right) = \frac{1}{2} \quad \text{if} \quad \left|\frac{y_j - Y_{ji}}{h_i}\right| \leqslant 1 \quad (5.2\text{-}13)$$

$i = 1, 2, \ldots, p$. In this case the density estimate is

$$f_n(\mathbf{y}) = \frac{\Delta_p F_n(\mathbf{y})}{2^p h_1 h_2 \cdots h_p} \quad (5.2\text{-}14)$$

where $\Delta_p F_n(\mathbf{y})$ is the empirical CDF at $\mathbf{y}$. If we assume, for example, that conditional on X being in state m, Y has density $f_{i,m}$ under Π_i, then assuming the same set of h_i values are used for both density estimates, the rule for classification becomes to assign $\mathbf{w} = (\mathbf{y}, \mathbf{x})$ to Π_1 if

$$\frac{\Delta_p F_{n_{1m}}(\mathbf{y})}{\Delta_p F_{n_{2m}}(\mathbf{y})} \geqslant \frac{p_{2m}}{p_{1m}} \quad (5.2\text{-}15)$$

where $F_{n_{im}}(\mathbf{y})$ is the empirical CDF in state m from population Π_i based on n_{im} sample observations.

5.3 METHODS FOR COMPARING PROCEDURES

With the rather large number of procedures discussed here, in addition to the standard LDF and its variants, it is of considerable interest to determine the relative effectiveness within a given data set of competing classification procedures. In various sampling studies, for example, those of Gilbert [1968], Moore [1973], and Dillon and Goldstein [1978] (discussed in Section 5.4), the authors attempt to associate underlying parametric structure in the two groups to the relative effectiveness of the procedures discussed. For the most part studies of this nature have been restricted to multivariate binary data, and even here general conclusions are most difficult. As a different approach to this type of analysis we discuss some methods considered by Goldstein [1975, 1976].

Suppose that P_1 and P_2 are two procedures that have been established on the basis of two independent sets of independent samples from $\Pi_1, \Pi_2, \ldots, \Pi_M$. Note that we depart from the two-population problem since the general case is as easy to discuss. Suppose, in addition, that a total of $n + m$ observations are available from $\Pi_i, i = 1, 2, \ldots, M$. We use this additional set of observations as test points, n of which are to be reclassified by P_1 and the remaining m by P_2. Let $n_i^1(m_i^1)$ be the number of observations of the available $n(m)$ that are known to belong to Π_i and are correctly classified. Let $n_i^{-1}(m_i^{-1})$ be the number of observations of the available $n(m)$ that are known to belong to Π_i and are incorrectly classified. Further, let $n_i = n_i^1 + n_i^{-1}$ and $m_i = m_i^1 + m_i^{-1}$. We consider two sampling situations:

1. The $n_i(m_i)$ are fixed and $n = \sum_{i=1}^M n_i, m = \sum_{i=1}^M m_i$
2. The $n_i(m_i)$ are randomly generated to the M populations with $\sum_{i=1}^M n_i = n, \sum_{i=1}^M m_i = m$ according to a uniform multinomial with each state probability equal to $1/M$.

Finally, suppose that $n_i^e(m_i^e)$ denotes the expected number of correctly classified observations from Π_i and $n_i^{-e}(m_i^{-e})$, the expected number of incorrectly classified observations from Π_i. The expected number of correctly or incorrectly classified observations take on different values depending on which sampling case we consider.

Consider first case 1. Let

$$I_{ij} = \begin{cases} 1 \text{ if the } j\text{th observation is from } \Pi_i \text{ and is correctly} \\ \quad \text{classified by random allocation;} \\ 0 \text{ otherwise.} \end{cases}$$

But $n_i^e = \sum_{j=1}^n E(I_{ij}) = (n_i/M)$ since $E(I_{ij}) = (n_i/n) \cdot (1/M)$, and hence

$$Q_1 = \sum_{i=1}^M \left[\left\{ \frac{(n_i^1 - n_i^e)^2}{n_i^e} \right\} + \left\{ \frac{(n_i^{-1} - n_i^{-e})^2}{n_i^{-e}} \right\} \right] \tag{5.3-1}$$

is a sum of M independent (asymptotically) χ^2 random variables each with 1 degree of freedom and hence is asymptotically χ^2 with M degrees of freedom. Similarly,

$$Q_2 = \sum_{i=1}^M \left[\left\{ \frac{(m_i^1 - m_i^e)^2}{m_i^e} \right\} + \left\{ \frac{(m_i^{-1} - m_i^{-e})^2}{m_i^{-e}} \right\} \right] \tag{5.3-2}$$

is asymptotically χ^2 with M degrees of freedom and is independent of Q_1. If $n_i^e = (n_i / M)$ and $m_i^e = (m_i / M)$ are substituted into the right-hand sides of (5.2-1) and (5.2-2), respectively, then

$$Q_1 = \sum_{i=1}^{M} \frac{\left(Mn_i^1 - n_i\right)^2}{\left[n_i(M-1)\right]}$$

and

$$Q_2 = \sum_{i=1}^{M} \frac{\left(Mm_i^1 - m_i\right)^2}{\left[m_i(M-1)\right]} . \tag{5.3-3}$$

Suppose we now consider the following hypothesis

$$H_0: P_1 = P_2$$

$$H_1: P_1 > P_2.$$

By $P_1 = P_2$ we mean that for both procedures the number of correct classifications took place at random, whereas by $P_1 > P_2$ we mean that procedure P_1 did better than P_2. Since Q_1 and Q_2 are independent, it follows under H_0 that

$$\frac{Q_1}{Q_2} = \frac{\left[\sum_{i=1}^{M} \left(Mn_i^1 - n_i\right)^2 / n_i\right]}{\left[\sum_{i=1}^{M} \left(Mm_i^1 - m_i\right)^2 / m_i\right]} \tag{5.3-4}$$

has an approximate asymptotic F distribution with M and M degrees of freedom.

Returning now to sampling plan 2, it follows that since $E(I_{ij}) = 1/M^2$ that $n_i^e = \sum_{j=1}^{n} E(I_{ij}) = (n/M^2)$ and similarly $m_i^e = (m/M^2)$. However, in this case the expressions in (5.2-1) and (5.2-2) have asymptotic χ^2 distribution with $2M-1$ degrees of freedom since now the $2 \times M$ contingency table has $2M-1$ independent cells. It follows, therefore, that when $n_i^e = (n/M^2)$ and $m_i^e = (m/M^2)$ are placed into the right-hand side of (5.2-1) and (5.2-2), respectively, and denoting these statistics by Q_1^* and Q_2^* we have

$$Q_1^* = \sum_{i=1}^{M} \left[(M-1)\left(M^2 n_i^1 - n\right)^2 + \left(M^2 n_i^{-1} - n(M-1)\right)\right]^2 \cdot \frac{1}{\left[nM^2(M-1)\right]}$$

$$\tag{5.3-5}$$

and

$$Q_2^* = \sum_{i=1}^{M} \left[(M-1)(M^2 m_i^1 - m)^2 + (M^2 m_i^{-1} - m(M-1)) \right]^2$$

$$\cdot \frac{1}{\left[mM^2(M-1) \right]} \cdot \qquad\qquad (5.3\text{-}6)$$

By the same reasoning, therefore, Q_1^*/Q_2^* is asymptotically F with $2M-1$ and $2M-1$ degrees of freedom under H_0.

If sufficient data are available, the ratios Q_1/Q_2 and Q_1^*/Q_2^* can be particularly helpful in determining whether a given competitor P_1 is really more useful than a standard procedure P_2. However, the authors have found, as probably should be expected, that significant differences only appear when the procedures involved are for a given data set very different. Perhaps some improvement in performance will result if instead of random assignment with uniform probabilities a weighting system for the probabilities is used.

5.4   SAMPLING EXPERIMENTS

Because of the large number of procedures available, it becomes of considerable interest to determine the relative effectiveness of competing classification rules. However, as we indicated earlier, there is a deficiency of adequate methodology for comparing alternative procedures. Hence the comparative studies to be reviewed here have been undertaken by means of Monte Carlo sampling experiments. These studies have attempted to determine whether linear procedures, such as Fisher's LDF, perform as well as multinomial classification rules on a variety of population structures characterized by underlying multivariate binary responses. In particular, attempts at general conclusions regarding the superiority of one procedure over another are made by associating the underlying parametric structure in the two groups to the relative effectiveness of the procedures.

The purpose of this section is to acquaint the reader with both the general framework taken and the comparative results reported in a number of Monte Carlo sampling studies. Although this section does not survey all such efforts, we discuss three well-referenced studies that all share the property of utilizing a Monte Carlo sampling framework to generate general conclusions as to the relative performance of a procedure. In the remainder of the section we discuss the different sampling schemes and reparametrizations used, the types of performance measures considered, some representative results, and general conclusions from the works of Gilbert [1968], Moore [1973], and Dillon and Goldstein [1978].

5.4.1   Sampling Schemes and Reparametrizations

Recall from our discussion of error rates in Chapter 3 that both the optimum and actual errors require that the underlying mass functions be known for their evaluation. From a theoretical persepective, therefore, the use of Monte Carlo sampling experiments is propitious to the extent that it allows one to determine the relative effectiveness of a procedure, since specification of the underlying parametric structure in the two groups is possible. However, in the case where the underlying multinomial distributions are assumed generated from multivariate binary data, specification of the parametric structure is most easily accomplished by use of a reparametrization of the state probabilities. For example, considerable simplification may be possible with the use of a reparametrization since certain higher-order terms can be set to zero, thereby reducing the number of parameters needed to describe the underlying distributions.

Two such reparametrizations have seen use, and we discussed both representations, at least in general terms, in Section 2.3. For completeness, the model reparametrizations utilized in the Monte Carlo sampling studies to be reviewed are sketched briefly in the following paragraphs.

The Loglinear Model.   This model representation was used by Gilbert [1968] to investigate the problem of classification for a variety of possible multivariate Bernoulli distributions. Recall from Section 2.3 that this representation expresses the $\log f(\mathbf{x})$ as a linear combination of main effects and interactions, namely:

$$\log f(\mathbf{x}) = \alpha + \sum_{j=1}^{p} (-1)^{x_j}\alpha_j + \sum_{j<k} (-1)^{x_j + x_k}\alpha_{jk} +$$

$$\vdots$$

$$+ (-1)^{x_1 + x_2 + \cdots + x_p}\alpha_{12\ldots p} \qquad (5.4\text{-}1)$$

where α is an overall effect, α_j is the main effect due to X_j, α_{jk} is the first-order interaction effect due to X_j and X_k, and so on. Expression (5.4-1) is a full representation in that $2^p - 1$ independent parameters are required to completely characterize the underlying distributions and, therefore, it is of limited use in Monte Carlo sampling since far too many parameters need estimation. However, by setting certain interaction terms to zero the number of parameters to be considered in the sampling experiments can be reduced to a smaller and more manageable set.

Indeed, in the Monte Carlo sampling experiments reported by Gilbert it was assumed that all second-and higher-order terms vanish so that

$$\log f_i(\mathbf{x}) = \alpha + \sum_{j=1}^{p} (-1)^{x_j} \alpha_j + \sum_{j<k} (-1)^{x_j + x_k} \alpha_{jk}. \qquad (5.4\text{-}2)$$

Note that in the above we let $X_{p+1} \equiv i$, and, therefore, the $f_i, i = 1, 2$ are the corresponding densities associated with populations Π_1 and Π_2, respectively.

The Bahadur Model. This model uses a reparametrization of $f_i(\mathbf{x})$ in terms of means and correlations. As we indicated in Section 2.3, Bahadur [1961] showed that the state probabilities of the 2^p states can be expressed in the following manner. Let

$$\theta_{ij} = E_i(X_j) \quad i = 1, 2, j = 1, 2, \ldots, p$$

$$Z_{ij} = \frac{(X_j - \theta_{ij})}{\left(\theta_{ij}(1 - \theta_{ij})\right)^{1/2}}$$

$$\rho_i(jk) = E(Z_{ij} Z_{ik})$$

$$\rho_i(jkl) = E(Z_{ij} Z_{ik} Z_{il}) \qquad (5.4\text{-}3)$$

$$(\text{etc.}).$$

Then

$$f_i(\mathbf{x}) = \prod_{j=1}^{p} \theta_{ij}^{x_j} (1 - \theta_{ij})^{1 - x_j}$$

$$\left[1 + \sum_{j<k} \rho_i(jk) Z_{ij} Z_{ik} + \sum_{j<k<l} \rho_i(jkl) Z_{ij} Z_{ik} Z_{il} \right.$$

$$\left. + \ldots + \rho_i(1, 2 \ldots p) Z_{i1} Z_{i2} \ldots Z_{ip} \right]. \qquad (5.4\text{-}4)$$

Note that we have again used the subscript $i, i = 1, 2$, to distinguish parameters under Π_1 as opposed to those from Π_2 and that $\rho_i(jk)$ is the ordinary product–moment correlation between variables X_j and X_k. Similar to the approach taken by Gilbert, both Moore [1973] and Dillon and Goldstein [1978] utilized an approximation to (5.4-4) to reduce the number of parameters to be estimated. Specifically, both studies considered samples from populations in which all correlation terms beyond first order were assumed zero. In this case the $f_i(\mathbf{x})$ are given by

$$f_i(\mathbf{x}) = \prod_{j=1}^{p} \theta_{ij}^{x_j}(1-\theta_{ij})^{1-x_j}\left\{1 + \sum_{j<k} \rho_{jk}Z_{ij}Z_{ik}\right\}. \qquad (5.4\text{-}5)$$

For the purpose of implementing Monte Carlo sampling experiments the Bahadur model appears to have several advantages over the log linear representation. The basic problem with the use of a first-order interaction model is that the generated populations are not described in terms of familiar parameters (i.e., means and correlations); thus, it is somewhat unnatural to determine for any particular data set whether the conclusions concerning the superiority of one procedure over another apply. In addition, the class of population structures that can be generated from a first-order interaction model is rather limited.

Once a reparametrization has been decided on, the method for generating Monte Carlo samples can be outlined as follows:

1. Parameter values, which define a particular population structure and provide the input values for the Monte Carlo sampling experiments, are specified and are used to determine the state probabilities.
2. Samples of a predetermined size are taken from the set of probability distributions by using a random number generator, such as the IBM SSP random number generator RANDU, and these samples are then used to determine the estimated state probabilities and the associated classification rule obtained from the use of a particular procedure.
3. Performance measures are determined for each procedure.
4. Steps 2 and 3 are repeated T times, and mean performance measures are computed on the basis of the T trials.

5.4.2 Measures of Performance

To effectively compare competing classification requires that measures of performance or worthiness be evaluated. From our discussion of error rates in Chapter 3, two likely candidates for assessing the relative effectiveness of a procedure are the actual and apparent errors. In keeping with the notation of Chapter 3, recall that a classification rule may be defined as an ordered partition $D = \langle D_1, D_2 \rangle$ of the sample space $\mathcal{X}$ having the property that any randomly drawn point $\mathbf{x}$ is allocated to Π_1 if and only if $\mathbf{x} \in D_i$. The conditional probability of misclassification given $\mathbf{x} \in D_j$ is given by $t(D|\mathbf{x}) = g_i(\mathbf{x})/g(\mathbf{x})$ for $i \neq j$, where $g_i(\mathbf{x})$ are the respective discriminant scores and $g(\mathbf{x})$ is the unconditional density at the point $\mathbf{x}$.

Therefore, the conditional probability of misclassification is

$$t(D) = E\{t(D|\mathbf{x})\} = \sum_{D_1} g_2(\mathbf{x}) + \sum_{D_2} g_1(\mathbf{x}). \qquad (5.4\text{-}6)$$

An optimal rule can be constructed by minimizing the conditional probability of misclassification at every point in $\mathfrak{X}$. Hence if we regard t as a function on the domain $\mathfrak{D}$ of all rules and define

$$t^* = \inf_{D \in \mathfrak{D}} t(D) \qquad (5.4\text{-}7)$$

then D is optimal if $t(D) = t^*$. Noting that an optimal partition D^* must be characterized by $\mathbf{x} \in D_1^*$ if $g_1(\mathbf{x}) > g_2(\mathbf{x})$ or $\mathbf{x} \in D_2^*$ if $g_1(\mathbf{x}) < g_2(\mathbf{x})$, and arbitrarily assigned if $g_1(\mathbf{x}) = g_2(\mathbf{x})$, it follows that

$$t^* = t(D^*) = \sum_{\mathbf{x}} \min(g_1(\mathbf{x}), g_2(\mathbf{x})). \qquad (5.4\text{-}8)$$

In practice we generally are not in a position to evaluate the optimal rule since we do not know the discriminant scores and thus our interest focuses on the performance of a sample-based rule $\hat{D}$. Recall, however, that the sample-based rule $\hat{D}$ generates two error rates—namely, the actual and apparent errors, which are in general different from t^*.

The Actual Error Rate. Suppose some sample-based procedure yielded the rule $\hat{D} = \langle \hat{D}_1, \hat{D}_2 \rangle$; then the value of the error-rate function at $\hat{D}, t(\hat{D})$, is called the actual error and assumes the form

$$\sum_{\hat{D}_1} g_2(\mathbf{x}) + \sum_{\hat{D}_2} g_1(\mathbf{x}). \qquad (5.4\text{-}9)$$

Note that the actual error, like the optimal error, is only of interest in a theoretical study since the discriminant scores need to be known for its evaluation.

The Apparent Error Rate. An intuitive estimate based on $\hat{D}$ of either the optimal error or the actual error is the apparent error, $\hat{t}(\hat{D})$, simply defined as the proportion of errors made by the rule. The apparent error assumes the form:

$$\hat{t}(\hat{D}) = \sum_{\mathbf{x}} \min \frac{\{N_1(\mathbf{x}), N_2(\mathbf{x})\}}{(N_1 + N_2)} \qquad (5.4\text{-}10)$$

where $N_i(\mathbf{x}), i = 1, 2$ is the number of sample values from population Π_1 and Π_2, respectively, with value $\mathbf{x}$, whereas $N_1 = \Sigma_\mathbf{x} N_1(\mathbf{x})$ and $N_2 = \Sigma_\mathbf{x} N_2(\mathbf{x})$ are the total samples from Π_1 and Π_2.

In Monte Carlo sampling experiments both the apparent and actual errors have been used to assess the performance of a procedure. However, for each of the procedures under consideration it is common practice to report its associated mean apparent and mean actual errors (i.e., the average error rate based on the number of trials initiated). In addition, since the apparent error is optimistically biased whereas the actual error is pessimistically biased, both of these performance measures are compared to the optimum error—the error rate obtained with the use of the optimal rule. Hence, two indicators, mean apparent error as compared to optimum error and mean increase in actual over optimum error, are commonly utilized for this purpose.

Another measure used in Monte Carlo sampling experiments to assess the relative effectiveness of a procedure is the mean correlation between the estimated likelihood ratio and true likelihood ratio. Letting $\hat{L}_j(\mathbf{x}) = \hat{P}_j\{\mathbf{X} = \mathbf{x}|\Pi_1\} / \hat{P}_j\{\mathbf{X} = \mathbf{x}|\Pi_2\}$ be the ratio of likelihoods obtained with the use of procedure j, and let $L^*(\mathbf{x})$ be the ratio of likelihoods based on the optimal rule D^*, then the correlation coefficient is given by

$$ r_j = \frac{\text{cov}\{L^*(\mathbf{x}), L_j(\mathbf{x})\}}{(\text{var}\, L^*(\mathbf{x})\, \text{var}\, \hat{L}_j(\mathbf{x}))^{1/2}} \tag{5.4-11} $$

where the variances and covariances are taken over the marginal distribution of $\mathbf{x}$.

5.4.3 Some Representative Results

In this section we present some representative results from the studies of Gilbert [1968], Moore [1973], and Dillon and Goldstein [1978]. As we indicated, the two latter studies were similar in that both used a second-order Bahadur approximation to generate the population structures and, in addition, some of the same procedures were evaluated. Therefore, we rely primarily on the results reported by the latter authors since a richer class of mean and correlation structures was considered.

The Gilbert Study [*1968*].   Gilbert studied the problem of classification for a variety of multivariate Bernoulli distributions generated by the first-order interaction model given in expression (5.4-2). In her study, five classification procedures were considered:

1. Full multinomial procedure (method M). The ratio $P\{X = x|\Pi_1\}/P\{X = x|\Pi_2\}$ is estimated by $N_1(x)/N_2(x)$, where $N_i(x)$, $i = 1, 2$, is the number of sample observations from population Π_i with $X = x$.

2. Independent-variable procedure (method I). Maximum likelihood estimates of the $P\{X = x|\Pi_i\}$, $i = 1, 2$ are obtained under the assumption that the variables are mutually independent in each of the two populations.

3. Maximum-likelihood procedure (method ML). Maximum-likelihood estimates of $\boldsymbol{\beta}$ are obtained under the assumption that

$$\log\left[\frac{P\{X = x|\Pi_1\}}{P\{X = x|\Pi_2\}}\right] = 2\boldsymbol{\beta}'\tilde{x} \qquad (5.4\text{-}12)$$

where $\boldsymbol{\beta}' = (\beta_0, \beta_1, \ldots, \beta_p)$ and $\tilde{x} = (1, x)$.

4. Minimum logit χ^2 procedure (method χ^2). Under the same setup as method ML, obtain the minimum logit χ^2 estimates of $\boldsymbol{\beta}$ that minimize

$$\sum \frac{N_1(x)N_2(x)}{N(x)}\left(\log\frac{N_1(x)}{N_2(x)} - 2\boldsymbol{\beta}'\tilde{x}\right)^2, \qquad (5.4\text{-}13)$$

where $N(x) = N_1(x) + N_2(x)$.

5. Linear discriminant function (method LDF). The last procedure investigated by Gilbert was Fisher's LDF.

Although method M has the advantage of being very simple to apply, it is likely to break down in practice if the number of variables is at all large, since many of the states will be too small to be effectively estimated with samples of reasonable size. Method I suffers from the generally untenable assumption of mutual independence. Methods ML and χ^2 fall between methods M and L with respect to their restrictiveness in that they allow some kinds of dependency. Assuming as a starting condition multivariate normality and identical covariance structures, method LDF is optimal in the sense of minimizing the probability of misclassification; however, in the case of binary variables it may not adequately form a basis for classification. Note that with the exception of method M, all procedures considered are linear models.

Monte Carlo sampling experiments were performed on 15 population structures with six variables. For each population structure 100 trials were initiated with $N = 100$ or $N = 500$. In this study N_i, $i = 1, 2$ was a random variable, and hence it is not possible to determine the prior probabilities. As was indicated, only a first-order interaction model was considered. The 15 population structures were generated from the following sets of parameter values:

β		α_j	
1.	(0.8, 0.8, 0.8, 0.8, 0.8, 0.8)	1.	(0.0, 0.0, 0.0, 0.0, 0.0, 0.0)
2.	(0.2, 0.5, 0.8, 1.1, 1.4, 1.7)	2.	(0.5, 0.5, 0.5, 0.5, 0.5, 0.5)
3.	(0.2, 0.2, 0.6, 0.6, 1.1, 1.1)	3.	(0.0, 0.2, 0.4, 0.6, 0.8, 1.0)
			as well as the permutation
4.	(0.3, 0.3, 0.3, 0.7, 0.7, 0.7)		(1.0, 0.8, 0.6, 0.4, 0.2, 0.0)
		4.	$(-.5,\ -.3,\ -.1,\ 0.1,\ 0.3,\ 0.5)$
			as well as various permutations

α_{jk}

1. $\alpha_{jk} = 0.0 \ (j, k = 1, 2, \ldots, 6)$
2. $\alpha_{12} = \alpha_{13} = \alpha_{23} = 0.4, \alpha_{45} = \alpha_{46} = \alpha_{56} = 0.4$, otherwise $\alpha_{jk} = 0.0$
3. $\alpha_{45} = \alpha_{46} = \alpha_{56} = 0.4$, otherwise $\alpha_{jk} = 0.0$

Four of the population structures (populations 7, 8, 13, and 14) were labeled by Gilbert as "extreme populations" because of state sparseness. With $N = 100$, estimation was a problem for methods ML and I in the extreme populations, and hence such cases were omitted in calculating the evaluation criteria.

The average correlation between estimated likelihood ratio and true likelihood ratio and the mean actual error were computed for each population structure and method. The results of the sampling experiments for the case where $N = 100$ are given in Table 5.4-1. The first figure for each method shows the mean actual error, whereas the second figure shows the mean correlation coefficient. With the exception of extreme populations, all of the linear methods consistently perform better than method M. Of the four linear methods, methods I and χ^2 generally perform better than methods ML and LDF; however, the differences are for the most part slight.

Based on the criteria of mean correlation and mean actual error, Gilbert concluded that for population structures considered it makes very little

TABLE 5.4-1

MEAN ACTUAL ERROR AND MEAN CORRELATION COEFFICIENT[a]

$N = 100$ METHOD[b]	POPULATION				
	1	2	3	4	5
OE	1766	1156	0721	2542	1152
M	2609	1676	1239	3405	1920
	8091	8242	8495	6587	8290
I	1782	1238	0796	2749	1225
	9844	9466	9823	9513	9903
ML	1812	1301	0845	2770	1305
	9713	9130	9635	9385	9762
χ^2	1796	1246	0897	2825	1321
	9778	9397	9598	9263	9751
LDF	1826	1292	0883	2763	1286
	9680	9247	9623	9392	9808
	6	7	8	9	10
OE	1038	0133	0472	1910	0950
M	1589	0373	0693	2426	1473
	8928	7678	8808	8538	8889
I	1094	c	____	1951	1006
	9925	____	____	9866	9906
ML	1159	____	____	2065	1106
	9705	____	____	9395	9560
χ^2	1122	0271	0531	2030	1030
	9849	8443	8641	9578	9738
LDF	1188	0247	0579	2047	1098
	9720	8193	8142	9465	9620
	11	12	13	14	15
OE	0413	0487	0578	1452	1568
M	0890	0775	0814	2154	2207
	9537	8794	7973	8765	8205
I	0437	____	____	1765	1651
	9940	____	____	9768	9802
ML	____	____	____	1647	1727
	____	____	____	9700	9420
χ^2	0466	0524	0653	1672	1670
	9913	9519	7894	9771	9622
LDF	0547	0602	0675	1677	1716
	9789	8041	6548	9719	9494

[a]Actual values multiplied by 10^4 and are based on 100 Monte Carlo trials with $N = 100$.

[b]Methods: OE—optimum error, M—full multinomial, I—independent variables, ML—maximum likelihood, χ^2—minimum logit χ^2, LDF—Fisher's LDF.

[c]Empty cells represent cases where more than 20% of the estimates do not exist.

difference which of the four linear techniques is used; however, all of these methods are at least slightly superior to method M. In particular, she recommends the use of Fisher's LDF since the sampling experiments suggested that the loss involved from its use as compared to the other methods is too small to be of much importance. In addition, she comments that as the number of variables increases Fisher's LDF should remain fairly stable and very likely would yield a classification rule superior to the other methods, with the possible exception of method I.

Perhaps the most serious limitation of the Gilbert study is that her conclusions are not made in terms of the parameter values α_j and α_{jk}, so that no guidelines are provided for deciding the conditions under which one method may provide better performance than another. With the exception of those population structures labeled as "extreme," no information is provided as to the parameter values characterizing the underlying distributions. In addition, considering that only a first-order interaction model was used to generate the populations, it is not surprising that the performance of the linear methods was generally better than that of method M.

The Dillon–Goldstein Study [1978]. The Monte Carlo sampling study reported by Dillon and Goldstein was based on population structures characterized by six variables, and in this sense is similar to the work of both Gilbert [1968] and Moore [1973]. However, rather than using a log linear representation to characterize the population structures, these authors used a second-order Bahadur approximation given in expression (5.4-5). For each sampling experiment 100 trials were performed. Population structures considered were characterized by the values assigned to the input parameters θ_{1j}, θ_{2j}, $\rho_1(jk)$, and $\rho_2(jk)$, where the first subscript in each case indicates the respective population. In addition to mean structures being characterized by marginal probabilities θ_{1j} and θ_{2j}, the authors considered structures determined by the difference $d_p = \theta_{2j} - \theta_{1j} \geqslant 0$; d_p was restricted to be less than or equal to 0.40 to make the study reasonable in size.

Three general classes of correlation structures were considered, and the results of each sampling experiment were reported in terms of either the difference between correlations, $d_p = \rho_{2j} - \rho_{1j}$, or by ρ, the common value assigned to all correlations.

Class (i) included structures in which all off-diagonal correlation terms in population Π_1 were set to zero, whereas a single nonzero correlation [$\rho_2(13)$] was specified in Π_2:

All $\rho_1(jk) = 0$ if $j \neq k$; all $\rho_2(jk) = 0$ if $j \neq k$ except $\rho_2(13)$

$= x$ for x between -0.60 and 0.60 such that d_ρ

$= -0.60, -0.40, -0.20, 0.00, 0.20, 0.40, 0.60.$

Class (ii) consisted of structures with nonzero off-diagonal correlations in both populations such that $\rho_1(jk) \neq \rho_2(jk)$ for all $j = k$. Two different sets were specified:

1. All $\rho_1(jk) = 0.10$ if $j \neq k$; all $\rho_2(jk) = x$ if $j \neq k$ for x between -0.05 and 0.33 such that $d_\rho = -0.15, 0.10, 0.00, 0.10, 0.20, 0.23$.
2. All $\rho_1(jk) = 0.20$ if $j \neq k$; all $\rho_2(jk) = x$ if $j \neq k$ for x between -0.05 and 0.33 such that $d_\rho = 0.25, -0.15, -0.10, 0.00, 0.10, 0.13$.

Class (iii) structures were restricted so that the correlation between any two variables in Π_1 was identical to that in Π_2. The case of equal correlation structure was considered since it allowed the investigation of situations in which the variance–covariance matrices in the two populations were identical. Note that when $\rho_1(jk) = \rho_2(jk)$ for all j and k, then letting $\theta_{2j} = 1 - \theta_{1j}$ for all j, yields equal covariance structures:

$$\text{All } \rho_1(jk) = \rho_2(jk) = \rho \text{ if } j \neq k \text{ with}$$

$$\rho = 0.00, 0.10, 0.20, 0.30, 0.33.$$

In addition to investigating three of the methods studied by Gilbert (methods M, I, and LDF), the authors evaluated three other procedures.

1. Second-order Procedure (Method S): This assumes a Bahadur reparametrization with correlations beyond those of the first-order zero. The remaining parameters are replaced by their maximum-likelihood estimates.

2. Martin–Bradley Procedures (Methods MB-C, MB-M, and MB-I): Recall from Section 2.3.3 that Martin and Bradley developed a class of probability models for multinomial distributions of the form

$$f_i(\mathbf{x}) = f(\mathbf{x})\left[1 + h(\mathbf{a}^{(i)}, \mathbf{x})\right] \tag{5.4-14}$$

for $i = 1, 2$, where $h(\mathbf{a}^{(i)}, \mathbf{x})$ is a polynomial in the elements of $\mathbf{x}$ and the coefficients $\mathbf{a}^{(i)}$ are specific to Π_i. The function $f(\mathbf{x})$ is defined by

$$f(\mathbf{x}) = w_1 f_1(\mathbf{x}) + w_2 f_2(\mathbf{x}), w_1 + w_2 = 1, w_i \geqslant 0, i = 1, 2 \tag{5.4-15}$$

where the weights w_i are regarded as arbitrary and assumed known or unknown (but estimated) depending on the sampling scheme. The $h(\mathbf{a}^{(i)}, \mathbf{x})$ are expressed in terms of orthogonal polynomials $\phi_\gamma(\mathbf{x})$, and in the complete expansion all polynomial terms up to and including order k are maintained. Assuming samples of size N_1 and N_2 taken from the two populations Π_1 and Π_2, respectively, the maximum likelihood estimators of $f_i(\mathbf{x})$, $i = 1, 2$ are given by

$$\hat{f}_i(\mathbf{x}) = \frac{N_i(\mathbf{x})}{N} \tag{5.4-16}$$

where $N = N_1 + N_2$ and $N_i(\mathbf{x})$ is again the number of sample observations from Π_i with $\mathbf{x}$. Further, $f(\mathbf{x})$ is estimated by

$$\hat{f}(\mathbf{x}) = \sum_{i=1}^{2} w_i \hat{f}_i(\mathbf{x}) \tag{5.4-17}$$

and $a_\gamma^{(i)}$ by

$$\hat{a}_\gamma^{(i)} = 2^{-k} \sum_{\mathbf{x}} \phi_\gamma(\mathbf{x}) Y_i(\mathbf{x}), \tag{5.4-18}$$

where $Y_i(\mathbf{x}) = \{\hat{f}_i(\mathbf{x}) - \hat{f}(\mathbf{x})\}/\hat{f}(\mathbf{x})$ provided $\hat{f}(\mathbf{x}) > 0$. Using these estimates for the unknown parameters yields the sample-based rule, which is to classify $\mathbf{x}$ into Π_1 if $h(\hat{\mathbf{a}}^{(1)}, \mathbf{x}) > h(\hat{\mathbf{a}}^{(2)}, \mathbf{x})$. Note that this rule is equivalent to the rule obtained with use of method M.

This classification rule is based on what Martin and Bradley call the "complete" model, and we refer to it as method MB-C. In addition to assessing the performance of this method, Dillon and Goldstein also considered two incomplete formulations:

1. Main effects procedure (method MB-M). Includes only main effect terms in the expansion of $h(\mathbf{a}^{(i)}, \mathbf{x})$.
2. First-order interaction procedure (method MB-I). Includes in the expansion of $h(\mathbf{a}^{(i)}, \mathbf{x})$ main effect and first-order interaction terms. As we indicated, if all first-order or second- and higher-order interaction terms are set to zero in $h(\mathbf{a}^{(i)}, \mathbf{x})$, then $f_i(\mathbf{x})$ may not be positive nor will probability sum to unity. However, it is nevertheless reasonable to examine whether the corresponding likelihood ratio rules provide an effective basis for classification.

3. Distance Procedure (method D): The distance procedure, which we discussed in Section 2.4, is different from other approaches in that the classification rule is formed not on the basis of likelihood ratios but on a measure of distributional distance. In keeping with the notation introduced in Section 2.4, suppose that $F_1 = \{p_j\}$ and $F_2 = \{q_j\}$, $j = 1, 2, \ldots, s$ are two discrete distributions defined on the same sample space. A measure of distance between F_1 and F_2 used to derive classification rules is given by

$$\|F_1 - F_2\|^2 = \sum_{j=1}^{s} \left(\sqrt{p_j} - \sqrt{q_j}\right)^2. \tag{5.4-19}$$

If n and m independent observations are taken from F_1 and F_2, respectively, let $S_n = \{n_i/n\}$ and $S_m = \{m_i/m\}$ be the derived empirical distributions. The rule for classifying a sample observation with $\mathbf{X} = \mathbf{x}$ is given by

Classify $\mathbf{X} = \mathbf{x}$ into F_1 if $\|S_{n+1} - S_m\| > \|S_n - S_{m+1}\|$,

Classify $\mathbf{X} = \mathbf{x}$ into F_2 if $\|S_{n+1} - S_m\| < \|S_n - S_{m+1}\|$,

Randomly allocate if $\|S_{n+1} - S_m\| = \|S_n - S_{m+1}\|$, $\qquad$ (5.4-20)

where S_{n+1} denotes the empirical distribution based on a sample of size $n+1$, similarly for S_{m+1}. Assuming equal prior probabilities and equal sample sizes $n = m$ the distance method is equivalent to method M. However, as we discussed earlier, if the sample sizes are not equal, then the two methods are different and hence their relative performance was assessed in situations of disproportionate sample sizes.

In this study, performance was primarily assessed on the basis of mean actual error and its increase over optimum error. The exceptions were for the Martin–Bradley models and the distance method. For the Martin–Bradley models (methods MB-C, MB-M, and MB-I), the mean apparent error as compared to optimum error was used to assess performance since the parameter structure characterizing this representation is not directly comparable to the Bahadur model, which is used to generate the Monte Carlo samples. In the case of the distance procedure (method D), the mean apparent, and mean actual errors were evaluated, and both were compared to the optimum error rate.

Tables 5.4-2 through 5.4-8 are adopted from Dillon and Goldstein [1978]. Table 5.4-2 presents the results of 100 Monte Carlo trials based on samples of size 200 and 400 under class (ii) correlation structures and mean structures as indicated in that table. With the exception of large absolute d_ρ terms, Table 5.4-2 reveals that the linear procedures (Methods F and LDF) do comparatively well. In situations of large absolute d_ρ, however, they do relatively poorly, especially compared to method S. This result is not too surprising in that the sampling experiments were initiated in the absence of third- and higher-order terms. Another interesting feature concerns the effect of equal but opposite "sign" d_ρ terms on discrimination. In particular, for method LDF it appears that negative correlations separate to a greater extent the two population groups than positive correlation. Recall that we commented on this result in Chapter 4 when discussing the variable selection problem.

Table 5.4-3 summarizes the results for class (ii) correlation structures with $d_\rho = 0.10$. Compared to methods M and S, the performance of the linear procedures is especially poor in situations of large absolute d_ρ. The great disparity between the linear procedures and the others can be

TABLE 5.4-2

MEAN INCREASE IN ACTUAL OVER OPTIMUM ERROR:
CLASS (I) CORRELATION STRUCTURES[a]

METHOD[b]	$d_\rho = \rho_2(13)$						
	−0.60	−0.40	−0.20	0.00	0.20	0.40	0.60
	$d_p = 0.10, n = m = 200$						
OE	3326	3711	3913	3997	3949	3753	3369
M	477	479	488	480	485	464	514
F	752	370	189	111	181	390	782
S	204	216	273	282	261	214	223
LDF	694	357	190	114	189	402	775
	$d_p = 0.10, n = m = 400$						
M	279	290	356	347	335	299	285
F	674	319	128	65	122	348	738
S	109	119	165	168	161	125	102
LDF	641	311	130	70	130	348	747
	$d_p = 0.30, n = m = 200$						
OE	1850	2008	2182	2174	2221	2209	2089
M	301	342	338	381	338	315	339
F	249	120	36	32	33	81	258
S	106	130	116	158	137	126	150
LDF	186	118	59	64	61	121	282
	$d_p = 0.30, n = m = 400$						
M	151	174	189	212	178	170	192
F	206	84	11	8	7	61	234
S	47	55	45	77	65	57	94
LDF	113	71	20	29	27	102	285

[a]Actual values multiplied by 10^4 and are based on 100 Monte Carlo trials.
[b]Methods: OE—optimum error, M—full multinomial, F—first-order Bahadur, S—second-order Bahadur, LDF—Fisher's linear discriminant function.

TABLE 5.4-3

MEAN INCREASE IN ACTUAL OVER OPTIMUM ERROR: CLASS (II) CORRELATION STRUCTURES[a]

$d_\rho = \rho_2(jk) - \rho_1(jk)$

METHOD[b]	All $\rho_1(jk)=0.10$							All $\rho_1(jk)=0.20$						
	-0.15	-0.10	-0.05	0.00	0.10	0.20	0.23	-0.25	-0.20	-0.15	-0.10	0.00	0.10	0.13
							$d_p=0.10, n=m=200$							
OE	3752	3992	4114	4232	4054	3736	3575	3133	3484	3779	3954	4240	4021	3951
M	417	405	404	346	428	322	298	500	525	452	478	299	283	243
F	352	173	105	42	320	747	945	1037	748	500	386	191	524	622
S	212	251	264	221	296	191	188	265	300	287	315	186	183	135
LDF	387	219	153	92	341	720	899	1090	793	546	411	188	457	536
							$d_p=0.10, n=m=400$							
M	287	295	303	263	315	210	179	336	329	330	351	211	183	111
F	308	141	65	18	327	779	990	962	715	466	359	206	564	674
S	120	144	167	146	222	118	118	129	154	167	206	131	122	63
LDF	332	179	102	44	345	776	940	1055	764	510	385	174	459	541

Source. Dillon and Goldstein [1978].

[a] Actual values multiplied by 10^4 and are based on 100 Monte Carlo trials.

[b] Methods: OE—optimum error, M—full multinomial, F—first-order Bahadur, S—second-order Bahadur, LDF—Fisher's linear discriminant function.

TABLE 5.4-4

Mean Increase in Actual Over Optimum Error: Class (III) Correlation Structures[a]

METHOD[b]	$\rho_1(jk)=\rho_2(jk)=\rho$									
	0.00	0.10	0.20	0.30	0.33	0.00	0.10	0.20	0.30	0.33
	$d_p=0.10, n=m=200$					$d_p=0.40, n=m=200$				
OE	3997	4232	4240	3863	3745	1390	2043	2363	1786	1349
M	486	346	299	427	455	330	290	320	358	363
F	119	42	191	742	919	14	2	325	1547	2170
S	297	221	188	310	322	114	150	238	242	239
LDF	130	92	188	614	749	30	35	276	1305	1846
	$d_p=0.10, n=m=400$					$d_p=0.40, n=m=400$				
M	362	263	211	346	333	159	181	217	231	179
F	67	18	206	776	970	1	0	332	1561	2195
S	177	146	131	221	195	3	89	134	134	111
LDF	74	44	174	631	760	3	10	302	1399	1952
	Identical Covariance Structure[c]									
	$d_p=0.20, n=m=200$					$d_p=0.20, n=m=400$				
OE	3174	3520	3578	2915	2716	3174	3520	3578	2915	2716
M	356	304	260	406	352	147	288	158	191	143
F	3	0	288	1296	1596	0	0	288	1296	1598
S	96	181	132	161	110	21	111	97	41	18
LDF	5	21	262	1119	1344	0	2	280	1215	1464

Source. Dillon and Goldstein [1978].

[a] Actual values multiplied by 10^4 and are based on 100 Monte Carlo trials.

[b] Methods: OE—optimum error, M—full multinomial, F—first-order Bahadur, S—second-order Bahadur, LDF—Fisher's linear discriminant function.

[c] When $\rho_1(jk)=\rho_2(jk)$ for all j and k, then letting $p_{2j}=1-p_{ij}$ for all j yields equal covariance structure. In our case, $p_{14}=0.4$, $p_{2j}=0.6$ for all j so that $d_p=0.20$.

TABLE 5.4-5

MEAN APPARENT ERROR FOR MARTIN–BRADLEY MODELS AND
SECOND-ORDER BAHADUR: CLASS (IIA) CORRELATION
STRUCTURES[a]

METHOD[b]	$d_p = \rho_2(jk) - \rho_1(jk)$						
	-0.15	-0.10	-0.05	0.00	0.10	0.20	0.23
	$d_p = 0.10, n = m = 200$						
OE	3752	3992	4114	4232	4054	3736	3575
MB-C(M)	2942	3091	3202	3231	3222	2984	2903
S	3392	3610	3712	3808	3725	3405	3272
MB-M	4241	4114	4040	4057	3953	3667	3457
MB-I	3410	3556	3678	3751	3632	3145	2981
	$d_p = 0.10, n = m = 400$						
MB-C(M)	3269	3465	3537	3579	3548	3284	3162
S	3562	3776	3878	3960	3877	3522	3398
MB-M	4318	4217	4153	4139	4012	3621	3375
MB-I	3592	3792	3884	3959	3840	3186	2910
	$d_p = 0.40, n = m = 200$						
OE	1527	1698	1870	2043	2389	2402	2367
MB-C(M)	1276	1421	1506	1615	1771	1863	1868
S	1458	1615	1738	1784	2093	2225	2204
MB-M	2026	2070	2169	2214	2439	2738	2876
MB-I	1774	1849	1930	1991	2331	2905	3069
	$d_p = 0.40, n = m = 400$						
MB-C(M)	1413	1562	1663	1783	1986	2044	2038
S	1498	1670	1792	1844	2209	2293	2273
MB-M	2003	2080	2159	2224	2408	2646	2854
MB-I	1801	1870	1960	2027	2344	2978	3180

Source. Dillon and Goldstein [1978].

[a]Actual values multiplied by 10^4 and are based on 100 Monte Carlo trials.

[b]Methods: OE—optimum error, MB-C—Martin–Bradley complete (equivalent to method M—full multinomial), S—second-order Bahadur, MB-M—Martin–Bradley main effects, MB-I—Martin–Bradley first-order interaction.

TABLE 5.4-6

MEAN APPARENT ERROR FOR MARTIN–BRADLEY MODELS AND SECOND-ORDER
BAHADUR: CLASS (III) CORRELATION STRUCTURES[a]

METHOD[b]	$\rho_1(jk)=\rho_2(jk)=\rho$									
	0.00	0.10	0.20	0.30	0.33		0.00	0.10	0.20	0.30	0.33
	$d_p=0.10, n=m=200$						$d_p=0.40, n=m=200$				
OE	3997	4232	4240	3863	3745		1390	2043	2363	1786	1349
MB-C(M)	3192	3231	3277	3251	3245		1207	1615	1780	1384	1162
S	3660	3808	3824	3664	3618		1340	1784	2177	1746	1301
MB-M	3923	4019	4246	4365	4348		2018	2223	2890	4247	4342
MB-I	3631	3720	3844	3797	3723		1811	1986	2933	2507	2116
	$d_p=0.10, n=m=400$						$d_p=0.40, n=m=400$				
MB-C(M)	3534	3579	3602	3551	3478		1323	1783	2007	1565	1255
S	3824	3960	3958	3815	3698		1379	1844	2253	1763	1330
MB-M	3953	4141	4356	4464	4354		2001	2238	2913	4341	4446
MB-I	3811	3980	4135	3907	3702		1813	2033	3033	2614	2141

Source. Dillon and Goldstein [1978].

[a] Actual values multiplied by 10^4 and are based on 100 Monte Carlo trials.
[b] Methods: OE—optimum error, MB-C—Martin–Bradley complete (equivalent to method M—full multinomial), S—second-order Bahadur, MB-M—Martin–Bradley main effects, MB-I—Martin–Bradley first-order interaction.

explained by the concept of a "reversal" in the sample-based likelihood ratios in that for certain population structures the true likelihood ratios do not increase monotonically, and hence the linear procedures cannot satisfactorily follow these reversals. The authors indicate, however, that when d_p was increased in size and d_ρ ranged between -0.15 and 0.20, the linear procedures did not perform as poorly; in fact, for $d_p=0.40$ all procedures did equally well.

The results for class (iii) correlation structures are shown in Table 5.4-4. Note that one population structure is presented in which the variance–covariance matrix in each population is identical, and we would thus expect method LDF to perform well in these samples. However, as before, the monotonicity of the likelihood ratios for the linear procedures results in very poor performance when large (in absolute value) correlations are present, even in situations where the samples were taken from populations having identical covariances. The authors also note that unlike the previous correlation structure discussed, the linear procedures fail to improve for large ρ when d_p is increased. Finally, note that the optimum error

TABLE 5.4-7

MEAN APPARENT AND ACTUAL ERRORS FOR UNEQUAL SAMPLE
SIZES: CLASS (IIA) CORRELATION STRUCTURES[a]

METHOD[b]	$d_\pi = \rho_2(jk) - \rho_1(jk)$						
	−0.15	−0.10	−0.05	0.00	0.10	0.20	0.23
	$d_p = 0.10, n = 50, m = 200$						
	Mean Apparent						
OE	3752	3992	4114	4232	4054	3736	3525
M	2181	2352	2363	2417	2435	2344	2300
D	2351	2408	2391	2445	2479	2539	2491
	Mean Actual						
M	4326	4528	4600	4649	4648	4345	4193
D	4191	4430	4644	4647	4524	4118	3750
	$d_p = 0.10, n = 50, m = 300$						
	Mean Apparent						
M	2242	2338	2441	2505	2486	2402	2401
D	2446	2415	2593	2564	2570	2668	2700
	Mean Actual						
M	4295	4490	4600	4685	4661	4420	4243
D	3910	4353	4612	4689	4666	4197	3726
	$d_p = 0.10, n = 100, m = 600$						
	Mean Apparent						
M	2826	3005	3056	3121	3084	2842	2798
D	2968	3122	2973	3134	3108	3001	2989
	Mean Actual						
M	4176	4383	4500	4574	4496	4292	3996
D	4018	4377	4505	4567	4498	4031	3610

Source. Dillon and Goldstein [1978].

[a]Actual values multiplied by 10^4 and are based on 100 Monte
Carlo trials.

[b]Methods: OE–optimum error, M—full multinomial, D—distance.

117

TABLE 5.4-8

MEAN APPARENT AND ACTUAL ERRORS FOR UNEQUAL SAMPLE SIZES:
CLASS (III) CORRELATION STRUCTURES[a]

METHOD[b]	$\rho_1(jk) = \rho_2(jk) = \rho$				
	0.00	0.10	0.20	0.30	0.33
	$d_p = 0.40, n = 50, m = 200$				
	Mean Apparent				
OE	1390	2043	2363	1786	1349
M	867	1128	1251	1010	827
D	1042	1260	1376	1223	1033
	Mean Actual				
M	1896	2464	2774	2295	1973
D	1900	2466	2780	2021	1633
	$d_p = 0.40, n = 50, m = 300$				
	Mean Apparent				
M	934	1214	1297	991	856
D	1179	1438	1553	1193	1072
	Mean Actual				
M	1888	2424	2712	2303	2076
D	1900	2454	2721	2102	1627
	$d_p = 0.40, n = 100, m = 600$				
	Mean Apparent				
M	1161	1553	1663	1291	1076
D	1369	1747	1902	1441	1263
	Mean Actual				
M	1686	2246	2645	2179	1899
D	1700	2300	2650	2050	1606

Source. Dillon and Goldstein [1978].

[a] Actual values multiplied by 10^4 and are based on 100 Monte Carlo trials.
[b] Methods: OE—optimum error, M—full multinomial, D—distance.

decreases with large ρ. This suggests, at least for binary predictors, that correlated variables may discriminate better than uncorrelated variables.

Tables 5.4-5 and 5.4-6 summarize results under class (iia) and class (iii) correlation structures, respectively, for the Martin–Bradley models. For comparative purposes, the tables also present results for method S. The tables indicate that in population structures characterized by small $d_p (= 0.10)$ and moderate correlations (i.e., those having relatively "large" optimum error rates), method MB-M yields less biased mean apparent errors than do the other methods. In addition, in situations of small $d_p (= 0.10)$ but either negative d_ρ or large positive ρ, method MB-I is slightly superior to method S. However, interestingly enough, when the mean difference d_p is increased to 0.40, the incomplete models fail to improve and their performance runs contrary to that reported under situations of small d_p. In particular, note that when d_p is increased to 0.40 the advantage of using method MB-M in situations of small positive ρ is nullified and, in addition, there seems to be a tendency for method MB-I, which performed quite well in situations of small d_p and either large positive ρ or negative d_ρ, to do progressively worse as d_ρ increases.

The sampling experiments for method D were conducted under three different sets of disproportionate sample sizes for classes (iia) and (iii) correlation structures. For two of the sets $n = 50$, with $m = 200$ and 300. The third set considered larger sample sizes; namely, $n = 100$ and $m = 600$. The authors note that for the case where $n = 50$, sparseness is clearly a problem since with six variables there are $2^6 = 64$ response patterns.

Tables 5.4-7 and 5.4-8 summarize results for methods D and M. As a general statement, method D almost always does better than method M, and in cases where it does not, the difference in performance is usually small. Furthermore, in the presence of either large absolute d_ρ or large positive ρ, the superiority of method D over method M can be seen to be in many cases especially striking. The rather strong performance of method D (over method M) in situations of disproportionate sample sizes led the authors to suggest that further study of this procedure is warranted to generate more authoritative conclusions as to its usefulness.

5.4.4　General Conclusions

This section has reviewed three studies that have investigated the performance of several classification rules under a wide variety of population structures. Although the comparisons among the procedures have been approached by simulating population structures, it should not be difficult to extrapolate from the sampling experiments to real-data situations if practitioners first analyze their data structures. Moreover, although attempts at general conclusion are for the most part difficult, improvements in discrimination may be realized if the data to be analyzed have estimates of parameters in certain neighborhoods. In conclusion, then, the results reported in the last section indicate that:

1. If parameter estimates are moderate, then use of a linear model should give reasonable results. However, the performance of linear models is in general quite similar, with no particular procedure having clear superiority.

2. The presence of correlated variables can decrease the accuracy of classification drastically in the case of the linear models. In addition, both Moore [1973] and Dillon and Goldstein [1978] found that the first-order Bahadur and LDF procedures performed poorly whenever the mean vectors in the two populations were similar.

3. Although the second-order Bahadur, in general, performed better than the other procedures, it is advisable to note that the sampling experiments were initiated in the absence of third- and higher-order terms and, therefore, its superiority is likely to be somewhat overstated.

4. The Martin–Bradley main effects and first-order incomplete models were shown to give reasonable results in situations of small mean differences. And at least in the presence of large absolute correlations, the first-order incomplete model was superior to the second-order Bahadur.

5. The performance of the distance method in situations of disproportionate sample sizes suggests both theoretical and practical credibility. Although a more extensive study is needed for real authoritative conclusions with respect to its usefulness in the presence of sparse states, the evidence reported by Dillon and Goldstein [1978] seems to be most encouraging.

CHAPTER 6

Computer Programs

6.1 INTRODUCTION

In the hope that the subject matter presented in this text will gain wider acceptance and subsequent use by applied researchers, we have decided to include program listings for a number of procedures discussed in Chapters 2, 4, and 5. However, so as not to proliferate the size of the chapter, our discussion of each program is brief and, therefore, the prospective user should contact one author (W. R. Dillon) if any additional documentation is needed.

Basically, the programs can be grouped into two main sets according to the chapter in which the procedure was originally discussed. For example, the first set, consisting of three programs, makes operational the following classification models discussed in Chapter 2: (a) the nearest-neighbor rule of order 1 (contained in PROG. NNRULE), (b) the full multinomial rule, the first- and second-order Bahadur models, and (c) the distance procedure (all contained in PROG. DISCRM); and the Martin and Bradley [1972] orthogonal polynomial model (PROG. POLY). In DISCRM, subroutines are also included to compute the usual LDF in the special case of binary predictor variables and, in addition, to calculate the Goldstein [1975, 1976] F-statistic, discussed in Section 5.3, which compares the performance of two respective classification rules. The second set of programs makes operational the three variable-selection procedures illustrated in Section 4.4, namely, the variable-selection procedure of Goldstein and Rabinowitz [1975] (PROG. DISTANCE), and the two sequential-selection procedures due to Lachin [1973] (PROG. LACHIN) and Goldstein and Dillon [1977] (PROG. VARSEL). With respect to the Lachin program, we would again like to take the opportunity to thank Professor John M. Lachin, who graciously made available his program listing and documentation.

All of the programs are written in the Fortran-IV language and, therefore, can be easily adapted to most systems. The first set of programs should be viewed as a collection of subroutines that can be easily modified to handle a variety of data sets. However, it should be remembered that with increasing numbers of variables or for variables having levels more than a dichotomy, core requirements are likely to become excessive. Similarly, for the variable-selection programs it is reasonable to expect that any realistic application will require a rather large core allocation. Therefore, it is recommended that INTEGER statements be used wherever possible and appropriate. Finally, the prospective user should carefully examine each program to determine which IBM scientific subroutine package (SSP) routines are required.

6.2 PROG. NNRULE

Description: The following program computes the nearest-neighbor rule of order 1 and calculates the misclassification probabilities for a particular data set. The program accepts as input either the user's original data matrix (coded in 1, 2 form—corresponding to the levels of the variables) or the observed state frequency counts in the two respective groups. The user can vary the input format by changing those program statements marked by a double asterisk (**).

Card Preparation: In addition to the data set, the user must include two cards containing control information. The following cards are required for running PROG. NNRULE:

1. *Problem Card*

Columns	Contents
1–6	Alphanumeric problem name
7–8	Number of variables
9	1 if input is in frequency count form
	0 if input is the user's original data matrix—coded in 1, 2 form
10–11	Critical value for determining group membership (leave blank if column 13 contains a 1)

2. *Variable Name and Level Card*

Columns	Contents
1–2	Alphanumeric–first-variable code
3–5	Number of levels of first variable

Columns	Contents
6–7	Alphanumeric–second-variable code
8–10	Number of levels of second variable

$\vdots$

3. *Data Cards.* If the input is the user's original data matrix the first variable read corresponds to IG, which is the value of the group membership variable; all predictor variables must be coded in 1, 2 form.

Output: The output of NNRULE includes:

1. The number of variables and their levels along with the sample sizes and induced distributions for each group.
2. The state frequency counts for each group induced by considering nearest neighbors of order 1.
3. The misclassification probabilities in each group for the nearest-neighbor rule—order 1.

Uses and Limitations: In the listing that follows, the program is dimensioned to handle four binary variables. It is a simple matter, however, to increase the number of binary variables considered by changing those dimension statements marked by an asterisk (*).

```
SAMPLE INPUT AND OUTPUT FOR THE REIS AND SMITH DATA
ANALYSED IN EXAMPLE 2.3-4

SAMPLE INPUT
REIDATO31
Q1002Q2002Q3002
   19   29
   57   49
   29   27
   63   53
   24   43
   37   52
   42   30
   68   42
```

SAMPLE OUTPUT
1QUESTIONNAIRE RESULTS....REIDAT

0LEVELS OF FACTORS
   Q1   2
   Q2   2
   Q3   2
0SAMPLE SIZES   GROUP1=339.
                GROUP2=325.
                TOTAL =664.

| RESPONSE | | GROUP1 | | GROUP2 | |
PATTERN		COUNTS	FREQ.	COUNTS	FREQ.
000	1	19	.056	29	.089
100	2	57	.168	49	.151
010	3	29	.086	27	.083
110	4	63	.186	53	.163
001	5	24	.071	43	.132
101	6	37	.109	52	.160
011	7	42	.124	30	.092
111	8	68	.201	42	.129

NEAREST NEIGHBORS

| RESPONSE | | GROUP1 | | GROUP2 | |
PATTERN		COUNTS	FREQ.	COUNTS	FREQ.
000	1	129	.381	148	.455
100	2	176	.519	183	.563
010	3	153	.451	139	.428
110	4	217	.640	171	.526
001	5	122	.360	154	.474
101	6	186	.549	186	.572
011	7	163	.481	142	.437
111	8	210	.619	177	.545

1   APPARENT ERRORS NEAREST NEIGHBOR

PI= .5000
   P(2/1)   .4041
   P(1/2)   .4677
   TOTAL P   .4359

124

```
C PROG. NNRULE
C THIS PROGRAM COMPUTES THE NEAREST NEIGHBOR RULE OF ORDER 1
C AND CALCULATES THE APPARENT ERRORS
      PROGRAM DILLON(INPUT,TAPE5=INPUT,OUTPUT,TAPE6=OUTPUT)
      DIMENSION NF1(8),NF2(8),NNF1(8),NNF2(8),II(3),              *
     1HEAD(3),LEVEL(3),PF1(8),PF2(8),PNF1(8),PNF2(8),             *
     2ER(3),XLR(8)                                                *
      INTEGER CONFIG(8,3),DATA(8)                                 *
      COMMON CONFIG
7778  READ(5,7779)PR,PR1,K,IDD,IDG
7779  FORMAT(A4,A2,I2,I1,I2)
      WRITE(6,7780)PR,PR1
7780  FORMAT(26H1QUESTIONNAIRE RESULTS....A4,A2//)
      READ(5,7781)(HEAD(I),LEVEL(I),I=1,K)
7781  FORMAT(3(A2,I3))
      WRITE(6,7782)(HEAD(I),LEVEL(I),I=1,K)
7782  FORMAT(18HOLEVELS OF FACTORS/(3X,A2,7X,I4))
      NV=K
      NC=LEVEL(1)
      DO 7783 I=2,K
7783  NC=NC*LEVEL(I)
      IF(IDD.EQ.1)GO TO 1500
      DO 7784 I=1,NC
      NF1(I)=0
      NF2(I)=0
      DATA(I)=0
7784  CONTINUE
1126  FORMAT(1X,I1,10X,I1,1X,2I1,3X,I1)                           **
  77  READ(5,1126)IG,(II(I),I=1,K)
      IF(EOF(5))7786,1125
1125  DAT=0.0
      M=II(1)
      MM=1
      DO 7700 I=2,K
      MM=MM*LEVEL(I-1)
7700  M=M+MM*(II(I)-1)
      IF(IG.GE.IDG) GO TO 1106
1104  NF2(M)=NF2(M)+AMAX1(1.0,DAT)
      GO TO 77
1106  NF1(M)=NF1(M)+AMAX1(1.0,DAT)
      GO TO 77
1500  DO 1502 I=1,NC
      READ(5,1501)NF1(I),NF2(I)
      IF(EOF(5))7786,1502
1502  CONTINUE
1501  FORMAT(I3,1X,I3)                                            **
7786  NS1=0
      NS2=0
      DO 1512 I=1,NC
      NS1=NS1+NF1(I)
1512  NS2=NS2+NF2(I)
      NS=NS1+NS2
1101  FORMAT(13HOSAMPLE SIZES2X,7HGROUP1=,F4.0/15X,7HGROUP2=,F4.0/
     115X,7HTOTAL =,F4.0/)
      XNS1=NS1
      XNS2=NS2
      XNS=NS
      WRITE(6,1101)XNS1,XNS2,XNS
      DO 11 I=1,NC
      PF1(I)=NF1(I)/XNS1
      PF2(I)=NF2(I)/XNS2
  11  CONTINUE
      PI=.5
      CALL PATTN(K)
      WRITE(6,6)
   6  FORMAT(2X,8HRESPONSE,10X,6HGROUP1,15X,6HGROUP2/2X,
     17HPATTERN,7X,6HCOUNTS,2X,5HFREQ.,8X,6HCOUNTS,2X,5HFREQ./)
```

125

```
      DO 110 I=1,NC
110 WRITE(6,8)(CONFIG(I,J),J=1,K),I,NF1(I),PF1(I),NF2(I),PF2(I)
  8 FORMAT(6X,3I1,2X,I3,1X,I6,F7.3,8X,I6,F7.3)
C COMPUTES NEAREST NEIGHBORS-ORDER1
      DO 30 I=1,NC
      DO 35 J=1,K
 35 DATA(I)=DATA(I)+(CONFIG(I,J)*(10**(K-J)))
 30 CONTINUE
      DO 40 I=1,NC
      NNF1(I)=NF1(I)
      NNF2(I)=NF2(I)
      DO 40 KK=1,K
      IVAL=10**(KK-1)
      IDAT1=DATA(I)+IVAL
      IDAT2=DATA(I)-IVAL
      DO 40 L=1,NC
      IF(IDAT1.NE.DATA(L)) GO TO 41
      GO TO 42
 41 IF(IDAT2.NE.DATA(L)) GO TO 40
 42 CONTINUE
      NNF1(I)=NNF1(I)+NF1(L)
      NNF2(I)=NNF2(I)+NF2(L)
 40 CONTINUE
      DO 45 I=1,NC
      PNF1(I)=NNF1(I)/XNS1
 45 PNF2(I)=NNF2(I)/XNS2
      WRITE(6,47)
 47 FORMAT(//,16X,17HNEAREST NEIGHBORS,//)
      WRITE(6,6)
      DO 100 I=1,NC
100 WRITE(6,8)(CONFIG(I,J),J=1,K),I,NNF1(I),
    1PNF1(I),NNF2(I),PNF2(I)
C COMPUTES MISCLASSIFICATION PROBABILITIES
      DO 51 I=1,NC
      IF(PNF1(I))48,48,49
 48 PNF1(I)=.00001
 49 IF(PNF2(I))50,50,51
 50 PNF2(I)=.00001
 51 XLR(I)=ALOG(PNF2(I)/PNF1(I))
      QI=1.0-PI
      CRIT=ALOG(PI/QI)
      DO 54 I=1,3
 54 ER(I)=0.0
      DO 56 J=1,NC
      IF(XLR(J)-CRIT)58,59,60
 58 ER(2)=ER(2)+PF2(J)
      GO TO 56
 59 ER(2)=ER(2)+PI*PF2(J)
      ER(1)=ER(1)+QI*PF1(J)
      GO TO 56
 60 ER(1)=ER(1)+PF1(J)
 56 CONTINUE
      ER(3)=PI*ER(1)+QI*ER(2)
      WRITE(6,210)
210 FORMAT(//,*1*,5X,*APPARENT ERRORS NEAREST NEIGHBOR*//)
      WRITE(6,211)PI
211 FORMAT(1H ,*PI=*,F6.4)
      WRITE(6,212)ER(1)
212 FORMAT(3X,*P(2/1)*,3X,F10.4)
      WRITE(6,213)ER(2)
```

126

```
213 FORMAT(3X,*P(1/2)*,3X,F10.4)
    WRITE(6,214)ER(3)
214 FORMAT(2X,*TOTAL P*,3X,F10.4)
    STOP
    END
C COMPUTES RESPONSE PATTERNS
        SUBROUTINE PATTN(N)
        INTEGER CONFIG(8,3)
        COMMON CONFIG
        NN=N
        L=1
 20 J=1
    S=1
 10 CONFIG(J,NN)=0
    IF(J-(2**(N-L))*(2*S-1))1,2,1
  1 J=J+1
    GO TO 10
  2 J=J+1
  4 CONFIG(J,NN)=1
    IF(J-(2**(N-L))*(2*S))5,6,5
  5 J=J+1
    GO TO 4
  6 IF(S-2**(L-1))7,8,7
  7 S=S+1
    GO TO 1
  8 IF(L-N)11,12,11
 11 L=L+1
    NN=NN-1
    GO TO 20
 12 CONTINUE
    RETURN
    END
```

6.3   PROG. DISCRM

Description: The following program consists of a number of subroutines that compute the full multinomial rule, the first- and second-order Bahadur models, the distance method, and the LDF. In addition, it also computes the *F*-statistic for the test described in Section 5.3. The program accepts as input either the user's original data matrix (coded in 1, 2 form—corresponding to the levels of the variables), or the observed state frequency counts in the two respective groups. The users must change those program statements marked by a double asterisk (**) to agree with the actual input format.

Card Preparation: This is the same as that for PROG. NNRULE.

Output: The output of DISCRM consists of the following:

1. The number of variables and their levels, along with the sample sizes and induced distributions for each group.

128 Computer Programs

2. The misclassification probabilities in each group for each procedure.
3. The results of computing the F-statistic for each pair of procedures.

Uses and Limitations: The program is dimensioned to handle at most four binary variables. To increase (decrease) the number of variables to be considered, the user needs to make several program changes. First, of course, if the number of variables is more than four, those dimension statements marked by an asterisk (*) must be changed so that array sizes are not exceeded. Second, subroutines MULTI, DREST, and XFISH must be modified each time the number of variables under consideration is changed. For example, if only three variables are examined those program statements marked by the letter C would be removed. On the other hand, if we desired to analyze five predictors, two statements would be added in each of these subroutines immediately following the last (C) statement, in other words,

$$
\begin{array}{ll}
\text{DO 20 I5} = 1,2 & \\
\text{N(5)} = \text{I5} - 1 & \text{(SUBR. MULTI)} \\
\text{DO 10 I5} = 1,2 & \\
\text{N(5)} = \text{I5} - 1 & \text{(SUBR. DREST)} \\
\text{DO 20 I5} = 1,2 & \\
\text{N(5)} = \text{I5} - 1 & \text{(SUBR. XFISH)}
\end{array}
$$

SAMPLE INPUT AND OUTPUT FOR THE DATA ANALYSED IN
EXAMPLE 2.2-1.

SAMPLE INPUT
DASHAT041
Q1002Q2002Q3002Q4002
```
  17    6
  32    8
   3    6
   2    8
  12    5
  26   30
   8    6
  14   33
   5    3
   3    4
   3    4
   3    3
   4   11
  15   23
   2   22
   5   86
```

```
C PROG. DISCRM                                                    GOLDSTEIN-DILLON
C THIS PROGRAM COMPUTES THE FULL MULTINOMIAL MODEL, THE FIRST
C AND SECOND ORDER BAHADUR MODELS, THE DISTANCE MODEL, AND THE LDF.
C IN ADDITION, IT ALSO COMPUTES THE F-STATISTIC FOR THE TEST DESCRIBED
C IN SECTION 5.3.
      PROGRAM DILLON(INPUT,TAPE5=INPUT,OUTPUT,TAPE6=OUTPUT)
      DIMENSION TP1(16),TP2(16),XLR(5,16),ER(3,5),PF1(16),             *
     1PF2(16),NF1(16),NF2(16),APER(3,5),ACER(3,5),SLR(5,16),           *
     2COR(5),ACOR(5),P1(4),P2(4),RT1(4,4),RT2(4,4),ERRINC(5)           *
      DIMENSION XNC1(2),XNM1(2),XMC2(2),XMM2(2),R1(2),R2(2)
      DIMENSION FF(10),TT(3,5),COV1(4,4),COV2(4,4)                     *
      DIMENSION XMP1(16),XMP2(16),XNF1(16),XNF2(16)                    *
      DIMENSION HEAD(4),LEVEL(4),II(4)                                 *
      INTEGER CONFIG(16,4)                                             *
      COMMON CONFIG
 7778 READ(5,7779)PR,PR1,K,IDD,IDG
 7779 FORMAT(A4,A2,I2,I1,I2)
      WRITE(6,7780)PR,PR1
 7780 FORMAT(26H1QUESTIONNAIRE RESULTS....A4,A2//)
      READ(5,7781)(HEAD(I),LEVEL(I),I=1,K)
 7781 FORMAT(4(A2,I3))
      WRITE(6,7782)(HEAD(I),LEVEL(I),I=1,K)
 7782 FORMAT(18HOLEVELS OF FACTORS/(3X,A2,7X,I4))
      NV=K
      NC=LEVEL(1)
      DO 7783 I=2,K
 7783 NC=NC*LEVEL(I)
      KKK=46
      IF(IDD.EQ.1)GO TO 1500
      DO 7784 I=1,NC
      NF1(I)=0
      NF2(I)=0
 7784 CONTINUE
 1126 FORMAT(1X,I1,10X,I1,1X,2I1,3X,I1)                                **
   77 READ(5,1126)IG,(II(I),I=1,K)
      IF(EOF(5))7786,1125
 1125 DAT=0.0
      M=II(1)
      MM=1
      DO 7700 I=2,K
      MM=MM*LEVEL(I-1)
 7700 M=M+MM*(II(I)-1)
      IF(IG.GE.IDG) GO TO 1106
 1104 NF2(M)=NF2(M)+AMAX1(1.0,DAT)
      GO TO 77
 1106 NF1(M)=NF1(M)+AMAX1(1.0,DAT)
      GO TO 77
 1500 DO 1502 I=1,NC
      READ(5,1501)NF1(I),NF2(I)
      IF(EOF(5))7786,1502
 1502 CONTINUE
 1501 FORMAT(I3,1X,I3)                                                 **
 7786 NS1=0
      NS2=0
      DO 1512 I=1,NC
      NS1=NS1+NF1(I)
 1512 NS2=NS2+NF2(I)
      NS=NS1+NS2
 1101 FORMAT(///,13HOSAMPLE SIZES2X,7HGROUP1=,F4.0/15X,7HGROUP2=,F4.0/
     115X,7HTOTAL =,F4.0//)
```

130

```
      WRITE(6,1101)XNS1,XNS2,XNS
      DO 11 I=1,NC
      PF1(I)=NF1(I)/XNS1
   11 PF2(I)=NF2(I)/XNS2
      CALL PATTN(K)
      WRITE(6,66)
   66 FORMAT(2X,8HRESPONSE,10X,6HGROUP1,15X,6HGROUP2/2X,
     17HPATTERN,7X,6HCOUNTS,2X,5HFREQ.,8X,6HCOUNTS,2X,5HFREQ./)
      DO 110 I=1,NC
  110 WRITE(6,8)(CONFIG(I,J),J=1,K),I,NF1(I),PF1(I),NF2(I),PF2(I)
    8 FORMAT(6X,4I1,2X,I3,1X,I6,F7.3,8X,I6,F7.3)
      LL=1
      MB1=0
      MB2=0
      DO 6 J=1,5
      DO 6 I=1,3
    6 APER(I,J)=0.0
      PI=XNS1/XNS
      CALL DXLR(NV,PF1,PF2,SLR,NC,NS1,NS2,NB1,NB2,LL,
     1XNS1,XNS2,NF1,NF2,P1,P2,RT1,RT2,COV1,COV2)
      MB1=MB1+NB1
      MB2=MB2+NB2
      CALL XMISS(PF1,PF2,SLR,ER,PI,NC)
      DO 15 I=1,3
      DO 15 J=1,5
   15 APER(I,J)=APER(I,J)+ER(I,J)
      WRITE(6,210)
  210 FORMAT(//15X,*APPARENT ERROR*//18X,*FULL*,5X,
     1*FIRST*,4X,*SECOND*,7X,*LDF*,2X,*DISTANCE*)
      CALL DOUT1(XNS1,XNS,APER)
      CALL FTEST(XNS1,XNS2,APER,FF,DF1,DF2)
      WRITE(6,600)
  600 FORMAT(//,15X,*F TEST RESULTS*//X,10HCOMPARISON,5X,*F CALCULATED*,
     15X,*D.F.*)
      WRITE(6,430)FF
  430 FORMAT(10HFULL-FIRST,9X,F5.3,9X,*3.    3.*/                         *
     111HFULL-SECOND,8X,F5.3,9X,*3.    3.*/                               *
     18HFULL-LDF,11X,F5.3,9X,*3.    3.*/13HFULL-DISTANCE,6X,F5.3,          *
     19X,*3.    3.*/12HFIRST-SECOND,7X,F5.3,9X,*3.    3.*/                 *
     19HFIRST-LDF,10X,F5.3,9X,*3.    3.*/14HFIRST-DISTANCE,5X,F5.3,        *
     19X,*3.    3.*/10HSECOND-LDF,9X,F5.3,9X,*3.    3.*/                   *
     115HSECOND-DISTANCE,4X,F5.3,9X,*3.    3.*/12HLDF-DISTANCE,7X,         *
     1F5.3,9X,*3.    3.*)                                                  *
      WRITE(6,203) MB1,MB2
  203 FORMAT(/*NO. OF TIMES SECOND ORDER PROBS.LT.ZERO*,2I5)
      STOP
      END
      SUBROUTINE DOUT1(XNS1,XNS,AMER)
C SUBROUTINE FOR PRINTING MISCLASSIFICATION PROBABILITIES              *
      DIMENSION AMER(3,5)
      PI=XNS1/XNS
      WRITE(6,211)PI
  211 FORMAT(/*PI=*,F6.4)
      WRITE(6,212)(AMER(1,I),I=1,5)
  212 FORMAT(3X*P(2/1)*,3X5F10.4)
      WRITE(6,213)(AMER(2,I),I=1,5)
  213 FORMAT(3X*P(1/2)*,3X5F10.4)
```

131

```
      WRITE(6,214)(AMER(3,I),I=1,5)                                    *
  214 FORMAT(2X*TOTAL P*,3X5F10.4)
      RETURN
      END
      SUBROUTINE DXLR(NV,TP1,TP2,XLR,NC,NS1,NS2,NB1,NB2,LL,XNS1,XNS2,
     1NF1,NF2,P1,P2,RT1,RT2,COV1,COV2)
C CALL VARIOUS SUBROUTINES
      DIMENSION TP1(16),TP2(16),XLR(5,16),P1(4),P2(4),                  *
     1RT1(4,4),RT2(4,4),PI1(16),PI2(16),PC1(16),PC2(16),               *
     2TLR(16),COV1(4,4),COV2(4,4),                                     *
     2NF1(16),NF2(16),XMP1(16),XMP2(16)                                *
      CALL DREST(NV,TP1,P1,RT1)
      CALL DREST(NV,TP2,P2,RT2)
      CALL MULTI(NV,P1,RT1,PI1,PC1,NB1)
      CALL MULTI(NV,P2,RT2,PI2,PC2,NB2)
      CALL LOGLK(TP1,TP2,TLR,NC)
      DO 10 J=1,NC
   10 XLR(1,J)=TLR(J)
      CALL LOGLK(PI1,PI2,TLR,NC)
      DO 11 J=1,NC
   11 XLR(2,J)=TLR(J)
      CALL LOGLK(PC1,PC2,TLR,NC)
      DO 12 J=1,NC
   12 XLR(3,J)=TLR(J)
      CALL CONCOR(P1,P2,RT1,RT2,COV1,COV2,NV)
      CALL XFISH(NV,NS1,NS2,COV1,COV2,P1,P2,TLR)
      DO 13 J=1,NC
   13 XLR(4,J)=TLR(J)
      CALL DSTANCE(XNS1,XNS2,NF1,NF2,NC,XMP1,XMP2,TLR)
      DO 14 J=1,NC
   14 XLR(5,J)=TLR(J)
   16 RETURN
      END
      SUBROUTINE XMISS(P1,P2,XLR,ER,PI,NC)
C COMPUTES MISCLASSIFICATION PROBABILITIES
      DIMENSION P1(16),P2(16),XLR(5,16),ER(3,5)                         *
      QI=1.0-PI
      CRIT=ALOG(PI/QI)
      DO 1 I=1,3
      DO 1 J=1,5
    1 ER(I,J)=0.0
      DO 10 I=1,5
      IF(I.EQ.5)CRIT=0.0
      DO 10 J=1,NC
      IF(XLR(I,J)-CRIT)5,6,7
    5 ER(2,I)=ER(2,I)+P2(J)
      GO TO 10
    6 ER(2,I)=ER(2,I)+PI*P2(J)
      ER(1,I)=ER(1,I)+QI*P1(J)
      GO TO 10
    7 ER(1,I)=ER(1,I)+P1(J)
   10 CONTINUE
      DO 15 I=1,5
   15 ER(3,I)=PI*ER(1,I)+QI*ER(2,I)
      RETURN
      END
C COMPUTES FIRST AND SECOND ORDER BAHADUR PROBABILITIES
      SUBROUTINE MULTI(NV,P,R,PI1,PI2,NB)
      DIMENSION P(4),PI1(16),PI2(16),N(4),Q(4),R(4,4)                  *
      NB=0
```

132

```
      DO 1 I=1,NV
    1 Q(I)=1.0-P(I)
      L=1
      DO 20 I1=1,2
      N(1)=I1-1
      DO 20 I2=1,2
      N(2)=I2-1
      DO 20 I3=1,2
      N(3)=I3-1
      DO 20 I4=1,2                                      C
      N(4)=I4-1                                         C
      PI1(L)=1.0
      DO 9 K=1,NV
      IF(N(K))7,7,8
    7 PI1(L)=PI1(L)*Q(K)
      GO TO 9
    8 PI1(L)=PI1(L)*P(K)
    9 CONTINUE
      NV1=NV-1
      STERM=1.0
      DO 10 I=1,NV1
      J1=I+1
      DO 10 J=J1,NV
      DENOM=P(I)*Q(I)*P(J)*Q(J)
      IF(DENOM.EQ.0.0)GO TO 11
      Z=(N(I)-P(I))*(N(J)-P(J))/SQRT(DENOM)
      GO TO 10
   11 Z=1.0
   10 STERM=STERM+Z*R(I,J)
      PI2(L)=PI1(L)*STERM
      IF(PI2(L).LT.0.0)NB=NB+1
      L=L+1
   20 CONTINUE
      RETURN
      END
      SUBROUTINE DREST(NR,PF,P,R)
C COMPUTES SAMPLE CORRELATIONS                          *
      DIMENSION PF(16),P(4),R(4,4),N(4)
      DO 2 I=1,NR
    1 P(I)=0.0
      DO 2 J=1,NR
    2 R(I,J)=0.0
      L=1
      DO 10 I1=1,2
      N(1)=I1-1
      DO 10 I2=1,2
      N(2)=I2-1
      DO 10 I3=1,2
      N(3)=I3-1
      DO 10 I4=1,2                                      C
      N(4)=I4-1                                         C
      DO 5 I=1,NR
      P(I)=N(I)*PF(L)+P(I)
      DO 5 J=1,NR
      R(I,J)=N(I)*N(J)*PF(L)+R(I,J)
    5 CONTINUE
   10 L=L+1
      DO 12 I=1,NR
      DO 12 J=1,NR
      DENOM=P(I)*(1.0-P(I))*P(J)*(1.0-P(J))
```

```
      IF(DENOM.EQ.0.0)GO TO 11
      R(I,J)=(R(I,J)-P(I)*P(J))/SQRT(DENOM)
      GO TO 12
  11  R(I,J)=1.0
  12  CONTINUE
  22  FORMAT(1H ,*P=*,F15.7,2X,6(F10.5))
      RETURN
      END
      SUBROUTINE LOGLK(X1,X2,XLR,N)
C THIS SUBROUTINE CALCULATES THE LOG LIKELIHOOD RATIOS
      DIMENSION X1(16),X2(16),XLR(16)
  10  FORMAT(//,2X,*LOG LIKELIHOOD RATIO RESULTS*)
      DO 5 I=1,N
      IF(X1(I))2,2,3
   2  X1(I)=.00001
   3  IF(X2(I))4,4,5
   4  X2(I)=.00001
   5  XLR(I)=ALOG(X2(I)/X1(I))
   6  FORMAT(1H ,5X,F15.7,2X,F15.7)
   7  FORMAT(1H ,*XLR=*,F15.7)
      RETURN
      END
      SUBROUTINE XFISH(NR,NS1,NS2,COV1,COV2,P1,P2,XSCOR)
C THIS SUBROUTINE COMPUTES FISHER'S LDF IN THE SPECIAL
C CASE WHERE X(I) ARE DICHOTOMOUS
      DIMENSION COV1(4,4),COV2(4,4),P1(4),P2(4),RLAM(4),XSCOR(16)    *
      DIMENSION D(4),S(4,4),LV(4),MV(4),N(4)                         *
      XNS=NS1+NS2
      PI1=NS1/XNS
      PI2=NS2/XNS
      DO 10 I=1,NR
  10  D(I)=P2(I)-P1(I)
      DO 12 I=1,NR
      DO 12 J=1,NR
  12  S(I,J)=PI1*COV1(I,J)+PI2*COV2(I,J)
      CALL MINV(S,NR,DET,LV,MV)
      DO 15 K=1,NR
      RLAM(K)=0.0
      DO 15 I=1,NR
  15  RLAM(K)=S(K,I)*D(I)+RLAM(K)
      CONST=0.0
      DO 16 K=1,NR
  16  CONST=CONST+RLAM(K)*(P1(K)+P2(K))
      CONST=0.5*CONST
      L=1
      DO 20 I1=1,2
      N(1)=I1-1
      DO 20 I2=1,2
      N(2)=I2-1
      DO 20 I3=1,2
      N(3)=I3-1
      DO 20 I4=1,2                                                   C
      N(4)=I4-1                                                      C
      XSCOR(L)=0.0
      DO 18 K=1,NR
  18  XSCOR(L)=RLAM(K)*N(K)+XSCOR(L)
      XSCOR(L)=XSCOR(L)-CONST
  20  L=L+1
      RETURN
      END
```

```
      SUBROUTINE CONCOR(P1,P2,RT1,RT2,COV1,COV2,NV)
C CALCULATES THE SAMPLE VARIANCE-COVARIANCE MATRICES
      DIMENSION P1(4),P2(4),RT1(4,4),RT2(4,4),COV1(4,4),COV2(4,4) *
      DO 5 I=1,NV
      DO 5 J=1,NV
      IF(I-J)2,3,4
    2 COV1(I,J)=RT1(I,J)*SQRT(P1(I)*P1(J)*(1.0-P1(I))*(1.0-P1(J)))
      COV2(I,J)=RT2(I,J)*SQRT(P2(I)*P2(J)*(1.0-P2(I))*(1.0-P2(J)))
      GO TO 5
    3 COV1(I,J)=P1(I)*(1.0-P1(I))
      COV2(I,J)=P2(I)*(1.0-P2(I))
      GO TO 5
    4 COV1(I,J)=COV1(J,I)
      COV2(I,J)=COV2(J,I)
    5 CONTINUE
      RETURN
      END
      SUBROUTINE FTEST(XNN,XMM,TT,FF,DF1,DF2)
C COMPUTES THE F STATISTIC FOR THE GOLDSTEIN TEST
C DESCRIBED IN SECTION 5.1
      DIMENSION XNC1(2),XNM1(2),XMC2(2),XMM2(2),R1(2),R2(2)       *
      DIMENSION FF(10),TT(3,5)                                    *
      J=1
      XM=2
      DO 100 K=1,4
      KK=K+1
      DO 90 K1=KK,5
      TOT1=0.0
      TOT2=0.0
      DO 10 I=1,2
      R1(I)=0.0
      R2(I)=0.0
      XNC1(I)=0.0
      XMC2(I)=0.0
      XNM1(I)=0.0
      XMM2(I)=0.0
   10 CONTINUE
      DO 11 I=1,2
      XNM1(I)=TT(I,K)*XNN
      XNC1(I)=XNN-XNM1(I)
      XMM2(I)=TT(I,K1)*XMM
      XMC2(I)=XMM-XMM2(I)
      T1=((XM**2)*XNC1(I))-XNN
      IF(T1)16,17,16
   16 R1(I)=(XM-1)*(T1**2)+((XM**2)*XNM1(I)-(XNN*(XM-1)))**2
   17 TOT1=TOT1+R1(I)
      T2=((XM**2)*XMC2(I))-XMM
      IF(T2)15,18,15
   15 R2(I)=(XM-1)*(T2**2)+((XM**2)*XMM2(I)-(XMM*(XM-1)))**2
   18 TOT2=TOT2+R2(I)
   11 CONTINUE
      FF(J)=(XMM*TOT1)/(XNN*TOT2)
      J=J+1
   90 CONTINUE
  100 CONTINUE
      DF1=(2*XM)-1
      DF2=(2*XM)-1
      RETURN
      END
      SUBROUTINE DSTANCE(XNS1,XNS2,NF1,NF2,N,XMP1,XMP2,TLR)
```

```
C COMPUTES THE DISTANCE MODEL
      DIMENSION XMP1(16),XMP2(16),NF1(16),NF2(16),
     1XNF1(16),XNF2(16),TLR(16)
      DO 9 I=1,N
      XNF1(I)=NF1(I)
      IF(XNF1(I).LT.0.0)XNF1(I)=0.0
      XNF2(I)=NF2(I)
      IF(XNF2(I).LT.0.0)XNF2(I)=0.0
    9 CONTINUE
      DO 10 K=1,N
      XMP1(K)=0.0
      XMP2(K)=0.0
   10 CONTINUE
      DO 100 J=1,N
      XMP1(J)=SQRT(((XNF1(J)+1)/(XNS1+1))*(XNF2(J)/XNS2))
      XMP2(J)=SQRT((XNF1(J)/XNS1)*((XNF2(J)+1)/(XNS2+1)))
      DO 99 I=1,N
      IF(I.EQ.J)GO TO 99
      XMP1(J)=XMP1(J)+SQRT((XNF1(I)/(XNS1+1))*(XNF2(I)/XNS2))
      XMP2(J)=XMP2(J)+SQRT((XNF1(I)/XNS1)*(XNF2(I)/(XNS2+1)))
   99 CONTINUE
  100 CONTINUE
      DO 110 I=1,N
  110 TLR(I)=ALOG(XMP1(I)/XMP2(I))
      RETURN
      END
C COMPUTES RESPONSE PATTERNS
      SUBROUTINE PATTN(N)
      INTEGER CONFIG(16,4)
      COMMON CONFIG
      NN=N
      L=1
   20 J=1
      S=1
   10 CONFIG(J,NN)=0
      IF(J-(2**(N-L))*(2*S-1))1,2,1
    1 J=J+1
      GO TO 10
    2 J=J+1
    4 CONFIG(J,NN)=1
      IF(J-(2**(N-L))*(2*S))5,6,5
    5 J=J+1
      GO TO 4
    6 IF(S-2**(L-1))7,8,7
    7 S=S+1
      GO TO 1
    8 IF(L-N)11,12,11
   11 L=L+1
      NN=NN-1
      GO TO 20
   12 CONTINUE
      RETURN
      END
```

```
SAMPLE OUTPUT
1QUESTIONNAIRE RESULTS....DASHAT


OLEVELS OF FACTORS
    Q1          2
    Q2          2
    Q3          2
    Q4          2


OSAMPLE SIZES   GROUP1=154.
                GROUP2=258.
                TOTAL =412.


    RESPONSE            GROUP1                  GROUP2
    PATTERN        COUNTS   FREQ.          COUNTS   FREQ.

       0000     1     17    .110              6     .023
       1000     2     32    .208              8     .031
       0100     3      3    .019              6     .023
       1100     4      2    .013              8     .031
       0010     5     12    .078              5     .019
       1010     6     26    .169             30     .116
       0110     7      8    .052              6     .023
       1110     8     14    .091             33     .128
       0001     9      5    .032              3     .012
       1001    10      3    .019              4     .016
       0101    11      3    .019              4     .016
       1101    12      3    .019              3     .012
       0011    13      4    .026             11     .043
       1011    14     15    .097             23     .089
       0111    15      2    .013             22     .085
       1111    16      5    .032             86     .333


                   APPARENT ERROR

                   FULL      FIRST    SECOND      LDF   DISTANCE

PI= .3738
    P(2/1)        .5122     .3506     .3312     .3636     .1948
    P(1/2)        .1129     .2713     .2403     .2403     .3566
    TOTAL P       .2621     .3010     .2743     .2864     .2961


                   F TEST RESULTS

    COMPARISON      F CALCULATED     D.F.
    FULL-FIRST          .822        3.      3.
    FULL-SECOND         .743        3.      3.
    FULL-LDF            .778        3.      3.
    FULL-DISTANCE       .688        3.      3.
    FIRST-SECOND        .539        3.      3.
    FIRST-LDF           .565        3.      3.
    FIRST-DISTANCE      .499        3.      3.
    SECOND-LDF          .625        3.      3.
    SECOND-DISTANCE     .552        3.      3.
    LDF-DISTANCE        .528        3.      3.

    NO. OF TIMES SECOND ORDER PROBS.LT.ZERO     0      0
```

137

6.4 PROG. POLY

Description: This program computes the Martin–Bradley orthogonal polynomial model. In addition to computing the complete model, it also reports results for the case where no more than main effects are used and when terms no more than first-order interactions are assumed. The program accepts as input either the user's original data matrix (coded in 1, 2 form—corresponding to the levels of the variables) or the observed state frequency counts across the two groups. The user can vary the input format by changing those program statements marked by a double asterisk (**).

Card Preparation: Two cards containing control information must be inputed along with the data set in order to execute POLY.

1. *Problem Card*

Columns	Contents
1–6	Alphanumeric problem name
7–8	Number of variables
9	1 if input is in frequency count form
	0 if input is the user's original data matrix—coded in 1, 2 form
10–11	Critical value for determining group membership (leave blank if column 13 contains a 1).
12–13	Number of coefficients to be computed—main effects and first-order interactions

2. *Variable Name and Label Card*

Columns	Contents
1–2	Alphanumeric—first-variable code
3–5	Number of levels of first variable
6–7	Alphanumeric—second-variable code
8–10	Number of levels of second variable
⋮	

3. *Data Cards*. If the input is the user's original data matrix, the first variable read corresponds to IG, which is the value of the group membership variable; all predictor variables must be coded in 1, 2 form.

Output: The output of POLY includes:

1. The number of variables and their levels along with the sample sizes, the induced distributions for each group and the estimates of the unconditional density at each **x**.
2. The misclassification probabilities in each group for the complete model, the main effects model, and the first-order interaction model.
3. The values of the $\hat{a}_\gamma^{(i)}$ coefficients that indicate the contribution of each variate when taken alone (main effect) and when combined with other variables (first-order interaction effects).

Uses and Limitations: In the listing that follows, the program is dimensioned to handle three binary variables. However, it is rather easy to increase the number of variables to be considered by simply changing those dimension statements marked by an asterisk (*). Note, in addition, that for each variable added (deleted), two program statements must be added (deleted) in subroutine POLY1. The location where the additions and deletions occur are marked by the letter C. For example, if one less variable was considered, those statements marked by a C would be removed, whereas if one more variable was added, the two program statements listed as follows would need to be included.

```
DO 50 I = 1,2
DO 50 J = 1,2
DO 50 K = 1,2 C
DO 50 L = 1,2 (addition)
Q(II,2) = I
Q(II,3) = J
Q(II,4) = K   C
Q(II,5) = L   (addition)
```

```
SAMPLE INPUT AND OUTPUT FOR THE REIS AND SMITH DATA ANALYSED
IN EXAMPLE 2.3-4.

SAMPLE INPUT
REIDAT0310007
Q1002Q2002Q3002
  19   29
  57   49
  29   27
  63   53
  24   43
  37   52
  42   30
  68   42
```

```
C PROG. POLY                                    GOLDSTEIN-DILLON
C THIS PROGRAM COMPUTES THE MARTIN AND BRADLEY ORTHOGONAL POLYNOMIAL
C MODELS, AND THEIR CORRESPONDING MISCLASSIFICATION PROBABILITIES
      PROGRAM DILLON(INPUT,TAPE5=INPUT,OUTPUT,TAPE6=OUTPUT)
      DIMENSION PF1(8),PF2(8),NF1(8),NF2(8),XNF1(8),XNF2(8)            *
      DIMENSION AF1(8),AF2(8),AAS1(8),AAS2(8)                          *
      DIMENSION HEAD(3),LEVEL(3),II(3),FN1(8),FN2(8),F(8)             *
      INTEGER CONFIG(8,3),DATA(8)                                     *
      COMMON CONFIG
 7778 READ(5,7779)PR,PR1,K,IDD,IDG,NCOEF
 7779 FORMAT(A4,A2,I2,I1,I2,I2)
      WRITE(6,7780)PR,PR1
 7780 FORMAT(26H1QUESTIONNAIRE RESULTS....A4,A2//)
      READ(5,7781)(HEAD(I),LEVEL(I),I=1,K)
 7781 FORMAT(3(A2,I3))
      WRITE(6,7782)(HEAD(I),LEVEL(I),I=1,K)
 7782 FORMAT(18HOLEVELS OF FACTORS/(3X,A2,7X,I4))
      NV=K
      NC=LEVEL(1)
      DO 7783 I=2,K
 7783 NC=NC*LEVEL(I)
      KKK=NCOEF
      IF(IDD.EQ.1)GO TO 1500
      DO 7784 I=1,NC
      NF1(I)=0
      NF2(I)=0
 7784 CONTINUE
 1126 FORMAT(I2,3I1)                                                  **
 7785 READ(5,1126)IG,(II(I),I=1,K)
      IF(EOF(5))7786,1127
 1127 DAT=0.0
      M=II(1)
      MM=1
      DO 7700 I=2,K
      MM=MM*LEVEL(I-1)
 7700 M=M+MM*(II(I)-1)
      IF(IG.GE.IDG) GO TO 1106
 1104 NF2(M)=NF2(M)+AMAX1(1.0,DAT)
      GO TO 7785
 1106 NF1(M)=NF1(M)+AMAX1(1.0,DAT)
      GO TO 7785
 1500 DO 1502 I=1,NC
      READ(5,1501)NF1(I),NF2(I)
      IF(EOF(5))7786,1502
 1502 CONTINUE
 1501 FORMAT(I3,1X,I3)                                                **
 7786 NS1=0
      NS2=0
      DO 1512 I=1,NC
      NS1=NS1+NF1(I)
 1512 NS2=NS2+NF2(I)
      NS=NS1+NS2
 1101 FORMAT(//,13HOSAMPLE SIZES2X,7HGROUP1=,F4.0/15X,7HGROUP2=,F4.0/
     115X,7HTOTAL  =,F4.0//)
      XNS1=NS1
      XNS2=NS2
      XNS=NS
      WRITE(6,1101)XNS1,XNS2,XNS
      DO 11 I=1,NC
      PF1(I)=NF1(I)/XNS1
```

```
   11 PF2(I)=NF2(I)/XNS2
      LL=1
      MB1=0
      MB2=0
      PI=XNS1/XNS
      CALL FATTN(K)
      CALL POLY1(NV,NC,NF1,NF2,NS1,NS2,FN1,FN2,F,AF1,AF2,
     1RC,RCM,RCI,NCOEF)
      WRITE(6,66)
   66 FORMAT(2X,8HRESPONSE,10X,6HGROUP1,15X,6HGROUP2,15X,*F(I)*/2X,
     17HPATTERN,7X,6HCOUNTS,2X,5HFREQ.,8X,6HCOUNTS,2X,5HFREQ.,10X,
     28HWEIGHTED/)
      DO 110 I=1,NC
  110 WRITE(6,8)(CONFIG(I,J),J=1,K),I,NF1(I),PF1(I),NF2(I),PF2(I),
     1 F(I)
    8 FORMAT(6X,3I1,2X,I3,1X,I6,F7.3,8X,I6,F7.3,12X,F7.3)
      WRITE(6,1400)
 1400 FORMAT(///15X,*MARTIN AND BRADLEY MODELS*//18X,*COMPLETE*,
     16X,*MAIN*,2X,*INTERACT*)
      WRITE(6,1401)RC,RCM,RCI
 1401 FORMAT(2X,*APPARENT ERROR*,3F10.4,//,32X,*COEFFICIENTS*,//)
      DO 1402 I=1,KKK
 1402 WRITE(6,1403)AF1(I),AF2(I)
 1403 FORMAT(/26X,2F10.4)
      STOP
      END
      SUBROUTINE POLY1(NV,NC,N1,N2,NN1,NN2,FN1,FN2,F,AF1,AF2,
     1RC,RCM,RCI,NCOEF)
      DIMENSION N1(8),N2(8),XN1(8),XN2(8),FN1(8),FN2(8),              *
     1F(8),Y1(8),Y2(8),YFI1(8),YFI2(8),                              *
     2AA1(8),AA2(8),YYY1(8),YYY2(8),AF1(8),AF2(8)                    *
      INTEGER Q(8,4),Z(8)                                            *
      KKK=NCOEF
      XNC=NC
      RRC=.9999
      XNN1=NN1
      XNN2=NN2
      NB=NV+1
      W1=0.5
      W2=0.5
      XNN=XNN1+XNN2
      P1=XNN1/XNN
      P2=XNN2/XNN
      DO 10 I=1,NC
      XN1(I)=N1(I)
      XN2(I)=N2(I)
   10 CONTINUE
      DO 17 I=1,NC
      IF(XN1(I))11,11,12
   11 FN1(I)=0.0
      GO TO 14
   12 FN1(I)=XN1(I)/XNN1
   14 IF(XN2(I))15,15,16
   15 FN2(I)=0.0
      GO TO 17
   16 FN2(I)=XN2(I)/XNN2
   17 CONTINUE
      DO 25 I=1,NC
      F(I)=(W1*FN1(I))+(W2*FN2(I))
   25 CONTINUE
```

```
              DO 33 I=1,NC
              IF(F(I))30,30,31
       30 Y1(I)=0.0
              Y2(I)=0.0
              GO TO 33
       31 Y1(I)=(FN1(I)-F(I))/F(I)
              Y2(I)=(FN2(I)-F(I))/F(I)
       33 CONTINUE
              DO 40 I=1,NC
              IF(F(I))37,37,38
       37 Z(I)=0.0
              GO TO 40
       38 Z(I)=1.0
       40 CONTINUE
              DO 41 J=1,NC
       41 Q(J,1)=1
              II=1
              DO 50 I=1,2
              DO 50 J=1,2
              DO 50 K=1,2
              Q(II,2)=I                                    C
              Q(II,3)=J
              Q(II,4)=K
              II=II+1                                      C
       50 CONTINUE
              DO 55 I=1,NC
              DO 55 J=1,NB
              IF(Q(I,J).EQ.1)GO TO 51
              Q(I,J)=-1
              GO TO 55
       51 Q(I,J)=1
       55 CONTINUE
C FULL MODEL ESTIMATION
              DO 95 I=1,NB
              AA1(I)=0.0
              AA2(I)=0.0
       95 CONTINUE
              DO 8998 I=1,KKK
              AF1(I)=0.0
     8998 AF2(I)=0.0
              DO 2500 J=1,NC
              YYY1(J)=0.0
     2500 YYY2(J)=0.0
              DO 960 I=1,NB
              DO 96 J=1,NC
              AA1(I)=AA1(I)+(Q(J,I)*Y1(J))
              AA2(I)=AA2(I)+(Q(J,I)*Y2(J))
       96 CONTINUE
              AF1(I)=AA1(I)/XNC
              AF2(I)=AA2(I)/XNC
      960 CONTINUE
              H1=0.0
              H2=0.0
              DO 6002 J=1,NC
              H1=H1+(F(J)*Y1(J))
              H2=H2+(F(J)*Y2(J))
     6002 CONTINUE
              CALL RISK(NC,Y1,Y2,F,P1,P2,RM)
              RC=RM
C MAIN EFFECTS MODEL ESTIMATION
```

```
      DO 2501 J=1,NC
      DO 2501 I=1,NB
      YYY1(J)=YYY1(J)+(AF1(I)*Q(J,I))
      YYY2(J)=YYY2(J)+(AF2(I)*Q(J,I))
 2501 CONTINUE
      DO 7500 J=1,NC
      IF(YYY1(J).GT.-1)GO TO 7506
      YYY1(J)=-.99999
 7506 IF(YYY2(J).GT.-1)GO TO 7500
      YYY2(J)=-.99999
 7500 CONTINUE
      H1=0.0
      H2=0.0
      DO 6003 J=1,NC
      H1=H1+(F(J)*YYY1(J))
      H2=H2+(F(J)*YYY2(J))
 6003 CONTINUE
      CALL RISK(NC,YYY1,YYY2,F,P1,P2,RM)
      RCM=RM
C FIRST-ORDER INTERACTION MODEL ESTIMATION
      II=NB
      DO 7000 I=2,NV
      KK=I+1
      DO 7000 K=KK,NB
      II=II+1
      DO 500 J=1,NC
      AF1(II)=AF1(II)+(Q(J,I)*Q(J,K)*Y1(J))
      AF2(II)=AF2(II)+(Q(J,I)*Q(J,K)*Y2(J))
  500 CONTINUE
      AF1(II)=AF1(II)/XNC
      AF2(II)=AF2(II)/XNC
 7000 CONTINUE
      DO 7005 J=1,NC
      YFI1(J)=YYY1(J)
 7005 YFI2(J)=YYY2(J)
      DO 504 J=1,NC
      II=NB+1
      DO 505 I=2,NV
      LL=I+1
      DO 506 K=LL,NB
      YFI1(J)=YFI1(J)+(AF1(II)*Q(J,I)*Q(J,K))
      YFI2(J)=YFI2(J)+(AF2(II)*Q(J,I)*Q(J,K))
      II=II+1
  506 CONTINUE
  505 CONTINUE
      II=0
  504 CONTINUE
      DO 7009 J=1,NC
      IF(YFI1(J).GT.-1)GO TO 7010
      YFI1(J)=-.99999
 7010 IF(YFI2(J).GT.-1)GO TO 7009
      YFI2(J)=-.99999
 7009 CONTINUE
      H1=0.0
      H2=0.0
      DO 7011 J=1,NC
      H1=H1+(F(J)*YFI1(J))
      H2=H2+(F(J)*YFI2(J))
 7011 CONTINUE
      CALL RISK(NC,YFI1,YFI2,F,P1,P2,RM)
```

143

```
              RCI=RM
              RETURN
              END
C SUBROUTINE FOR RISK ESTIMATIONS
              SUBROUTINE RISK(NC,YY1,YY2,F,P1,P2,RM)
              DIMENSION YY1(8),YY2(8),YZ1(8),YZ2(8),F(8),
             1R1(8),R2(8),R(8)
              DO 110 I=1,NC
              YZ1(I)=1.0+YY1(I)
              YZ2(I)=1.0+YY2(I)
          110 CONTINUE
              RM=0.0
              DO 120 I=1,NC
              R1(I)=P1*YZ1(I)
              R2(I)=P2*YZ2(I)
          120 CONTINUE
              DO 140 I=1,NC
              IF(R1(I)-R2(I))131,131,132
          131 R(I)=R1(I)
              GO TO 140
          132 R(I)=R2(I)
          140 CONTINUE
              DO 150 I=1,NC
              RM=RM+(F(I)*R(I))
          150 CONTINUE
              RETURN
              END
C COMPUTES RESPONSE PATTERN
              SUBROUTINE PATTN(N)
              INTEGER CONFIG(8,3)
              COMMON CONFIG
              NN=N
              L=1
           20 J=1
              S=1
           10 CONFIG(J,NN)=0
              IF(J-(2**(N-L))*(2*S-1))1,2,1
            1 J=J+1
              GO TO 10
            2 J=J+1
            4 CONFIG(J,NN)=1
              IF(J-(2**(N-L))*(2*S))5,6,5
            5 J=J+1
              GO TO 4
            6 IF(S-2**(L-1))7,8,7
            7 S=S+1
              GO TO 1
            8 IF(L-N)11,12,11
           11 L=L+1
              NN=NN-1
              GO TO 20
           12 CONTINUE
              RETURN
              END
```

144

```
SAMPLE OUTPUT
1QUESTIONNAIRE RESULTS....REIDAT


OLEVELS OF FACTORS
    Q1              2
    Q2              2
    Q3              2


OSAMPLE SIZES   GROUP1=339.
                GROUP2=325.
                TOTAL =664.


  RESPONSE              GROUP1              GROUP2              F(I)
  PATTERN          COUNTS  FREQ.       COUNTS  FREQ.         WEIGHTEI

     000      1      19    .056         29    .089            .073
     100      2      57    .168         49    .151            .159
     010      3      29    .086         27    .083            .084
     110      4      63    .186         53    .163            .174
     001      5      24    .071         43    .132            .102
     101      6      37    .109         52    .160            .135
     011      7      42    .124         30    .092            .108
     111      8      68    .201         42    .129            .165


              MARTIN AND BRADLEY MODELS

                     COMPLETE     MAIN   INTERACT
    APPARENT ERROR    .4232      .4301     .4259

                          COEFFICIENTS


                       -.0279      .0279

                        .0044     -.0044

                       -.1385      .1385

                       -.0647      .0647

                        .0751     -.0751

                       -.0187      .0187

                       -.0345      .0345
```

145

6.5 PROG. DISTANCE

Description: This is a variable-selection program that utilizes a pseudo distance between discriminant scores to find a "good" subset of variables. The selection process begins by computing $|(N_1(x_{(i)}))^{\frac{1}{2}} - (N_2(x_{(i)}))^{\frac{1}{2}}|$ for each subset of variables and then divides this (difference) by the appropriate scaling factor. Next, within each subset of variables, the program determines the scaled minimum and scaled average values of $|(N_1(x_{(i)}))^{\frac{1}{2}} - (N_2(x_{(i)}))^{\frac{1}{2}}|$ and then selects the maximum for subsets of a given size. Input to the program is the user's original data matrix coded in $1,2,\ldots,k$ form, where k corresponds to the maximum level for each variable, as well as some additional information.

Card Preparation: The following cards are mandatory for running DISTANCE:

1. *Problem Card*

Columns	Contents
1–6	Alphanumeric problem name
7–8	Number of variables
9–12	Blanks
13–14	Critical value for determining group membership

2. *Variable Name and Level Card*

Columns	Contents
1–2	Alphanumeric—first-variable code
3–5	Number of levels of first variable
6–7	Alphanumeric—second-variable code
8–10	Number of levels of second variable
⋮	

3. *Variable Format Card.* Using all 80 columns, specify the location of each variable in I format. Note that the first variable read corresponds to IG, which is the value of the group membership variable.
4. *Data Cards.* The data deck must be punched according to the variable format specified by the VARIABLE FORMAT card. Note the predictor variable must be coded in $1, 2,\ldots,k$ form.

Output: The output of DISTANCE consists of the following:

1. The number of variables and their levels, along with the sample sizes and induced distributions for each group.

2. The estimated scaled-minimum, scaled-sum, and scaled-average values of $\left|(N_1(x_{(i)}))^{\frac{1}{2}} - (N_2(x_{(i)}))^{\frac{1}{2}}\right|$ for all subsets of variables.
3. The number and percent misclassification obtained with the use of a particular subset of variables.
4. The maximum scaled minimum and scaled average of $\left|(N_1(x_{(i)}))^{\frac{1}{2}} - (N_2(x_{(i)}))^{\frac{1}{2}}\right|$ for each subset of size n.

Uses and Limitations: The program listing that follows is dimensioned to handle nine binary variables. However, the program is general in the sense that it can accommodate any number of discrete variables having k levels by simply changing those dimension statements marked by an asterisk (*). The method used for constructing arrays X and Y, which contain the original frequency counts and all lower-order sums in the two respective groups, is discussed by Hartley [1962] and further documentation can be found in subroutines AVDAT, AVCAL, and MEANQ of the IBM-SSP subroutine package. Note that the length of $X(Y)$ is given by $\Pi_j(L_j+1)$, where L_j is the number of levels of variable j. Therefore, whereas DISTANCE can easily be modified to handle any number of variables assuming k distinct levels, the prospective user should again be cautioned that core requirements can become excessive.

```
C PROG. DISTANCE                                          GOLDSTEIN-DILLON
C THIS PROGRAM COMPUTES THE MAXIMUM SCALED MINIMUM
C AND AVERAGE VALUES OF D FOR SUBSETS OF SIZE N.
      PROGRAM GR(INPUT,OUTPUT,TAPE5,TAPE6=OUTPUT,TAPE9)
      DIMENSION X(20000),Y(20000)
      INTEGER FINI,SH,PR,PR1,HEAD,FMT,BLANK,XDFNAME,SMNAME
      INTEGER TFM,CONFIG(512,9),XTOT,YTOT,TOTAL                        *
      DIMENSION TFM(15),II(9),XDFNAME(9,15),SIZE(9,2),SMNAME(9,15)     *
      DIMENSION HEAD(9),LEVEL(9),ISTEP(9),KOUNT(9),LASTS(9)           *
      DIMENSION SUMSQ(512),XDF(512),SMEAN(512),XMISS(512)             *
      DIMENSION FMT(15),PF1(512),PF2(512)                            *
    1 FORMAT(A4,A2,I2,A4,I2)
    2 FORMAT(1H1," QUESTIONAIRE RESULTS...."A4,A2//)
    3 FORMAT(18HOLEVELS OF FACTORS/(3X,A2,7X,I4))
    5 FORMAT(10HOSOURCE OF33X,7HMIN. OF10X,10HSCALED SUM9X,4HMEAN9X,
     X6HNUMBER8X,7HPERCENT/10H DISCRMNTN33X,
     1            7HSQRT. D11X,7HSQRT. D10X,7HSQRT. D5X,10HMISCLASSED5X,
     X10HMISCLASSED/)
    6 FORMAT(1H 15A2,2F19.5,F16.5,5X,F10.0,F15.2)
      DO 4 LOOP=1,K
      SIZE(LOOP,1)=SIZE(LOOP,2)=0.
      DO 4 L2=1,15
    4 XDFNAME(LOOP,L2)=SMNAME(LOOP,L2)=1H
  100 READ (5,1) PR,PR1,K,BLANK,IDG
      IF(EOF(5))9999,8888
 8888 DAT = 0
      WRITE (6,2) PR,PR1
      READ(5,7)                      (HEAD(I),LEVEL(I),I=1,K)
    7 FORMAT(9(A2,I3))
      WRITE (6,3) (HEAD(I),LEVEL(I),I=1,K)
      N=LEVEL(1)
      DO 102 I=2,K
  102 N=N*LEVEL(I)
      READ(5,521)TFM
  521 FORMAT(15A4)
      DO 101 I=1,20000
      X(I)=0.
  101 Y(I)=0.
  107 READ(9,TFM)IG,(II(I),I=1,K)
       IF (EOF(9)) 108,1077
 1077 M=II(1)
      MM=1
      DO 103 I=2,K
      MM=MM*LEVEL(I-1)
  103 M=M+MM*(II(I)-1)
      IF(IG.GE.IDG)GO TO 106
  104 X(M)=X(M)+AMAX1(1.0,DAT)
      GO TO 107
  106 Y(M)=Y(M)+AMAX1(1.0,DAT)
      GO TO 107
  108 DO 510 I=1,N
      XTOT=XTOT+X(I)
  510 YTOT=YTOT+Y(I)
      TOTAL=XTOT+YTOT
      DO 146 I=1,N
      PF1(I)=X(I)/XTOT
  146 PF2(I)=Y(I)/YTOT

      WRITE(6,511)XTOT,YTOT,TOTAL
  511 FORMAT(10(/),5X,"SAMPLE SIZE FOR G1 IS...",I7,/,
```

```
       A5X,"SAMPLE SIZE FOR G2 IS...",I7,/,
       B5X,"TOTAL SAMPLE SIZE   IS...",I7)
       CALL PATTN(K,CONFIG)
       WRITE(6,514)
 514 FORMAT(1H1,2X,8HRESPONSE,10X,6HGROUP1,15X,6HGROUP2/,3X,7HPATTERN,
     A7X,6HCOUNTS,2X,5HFREQ.,8X,6HCOUNTS,2X,5HFREQ./)
       DO 111 I=1,N
 111 WRITE(6,112)(CONFIG(I,J),J=1,K),X(I),PF1(I),Y(I),PF2(I)
 112 FORMAT(6X,9I1,3X,F3.0,5X,F4.3,8X,F3.0,2X,F4.3)                    *
       CALL AVDAT (K,LEVEL,N,X,L,ISTEP,KOUNT,Y)
       CALL AVCAL (K,LEVEL,X,L,ISTEP,LASTS,Y)
       CALL MEANQ (K,LEVEL,X,GMEAN,SUMSQ,XDF,SMEAN,ISTEP,KOUNT,LASTS,Y,
     1XMISS)
       WRITE (6,5)
       LL=(2**K)-1
       ISTEP(1)=1
       DO 105 I=2,K
 105 ISTEP(I)=0
       DO 110 I=1,15
 110 FMT(I)=BLANK
       NN=0
       SUM=0.0
 120 NN=NN+1
       L=0
       DO 140 I=1,K
       FMT(I)=BLANK
       IF(ISTEP(I)) 130, 140, 130
 130 L=L+1
       FMT(L)=HEAD(I)
 140 CONTINUE
       PCT=XMISS(NN)*100.0/GMEAN
       WRITE (6,6) (FMT(I),I=1,15),XDF(NN),SUMSQ(NN),SMEAN(NN),
     1XMISS(NN),PCT
       SUM=SUM+SUMSQ(NN)
       DO 10 L1=1,15
       IF(FMT(L1).NE.BLANK)GO TO 10
       IF(XDF(NN).LE.SIZE(L1-1,1)) 11,12
  11 IF(SMEAN(NN).LE.SIZE(L1-1,2))16,14
  12 SIZE(L1-1,1)=XDF(NN)
       DO 20 L2=1,K
  20 XDFNAME(L1-1,L2)=FMT(L2)
       GO TO 11
  14 SIZE(L1-1,2)=SMEAN(NN)
       DO 15 L3=1,K
  15 SMNAME(L1-1,L3)=FMT(L3)
       GO TO 16
  10 CONTINUE
  16 IF(NN-LL) 145, 170, 170
 145 DO 160 I=1,K
       IF(ISTEP(I)) 147, 150, 147
 147 ISTEP(I)=0
       GO TO 160
 150 ISTEP(I)=1
       GO TO 120
 160 CONTINUE
 170 N=N-1
       WRITE(6,997)
 997 FORMAT(1H1,10X,"MAXIMUMS FROM EACH NUMBER OF VARIABLES GROUPED",
     A 10(/),10X,"SOURCE",25X,"MAX OF AVERAGES",15X,
     B "SOURCE",25X,"MAX OF SCALED MINS")
```

```
      DO 1111 LOUT=1,K
1111  WRITE(6,998)(SMNAME(LOUT,I),I=1,K),SIZE(LOUT,2),
    A (XDFNAME(LOUT,I),I=1,K),SIZE(LOUT,1)
 998  FORMAT(10X,9A2,14X,F10.5,19X,9A2,14X,F10.5)          *
      GO TO 100
9999  STOP
      END
      SUBROUTINE PATTN(N,CONFIG)
      INTEGER CONFIG(512,9)
      NN=N
      L=1
  20  J=1
      S=1
  10  CONFIG(J,NN)=0
      IF(J-(2**(N-L))*(2*S-1))1,2,1
   1  J=J+1
      GO TO 10
   2  J=J+1
   4  CONFIG(J,NN)=1
      IF(J-(2**(N-L))*(2*S))5,6,5
   5  J=J+1
      GO TO 4
   6  IF(S-2**(L-1))7,8,7
   7  S=S+1
      GO TO 1
   8  IF(L-N)11,12,11
  11  L=L+1
      NN=NN-1
      GO TO 20
  12  CONTINUE
      RETURN
      END
      SUBROUTINE MEANQ (K,LEVEL,X,GMEAN,SUMSQ,XDF,SMEAN,MSTEP,KOUNT,
    1                  LASTS,Y,XMISS)
      DIMENSION LEVEL(9),X(20000),SUMSQ(512),XDF(512),SMEAN(512),
    1MSTEP(9),KOUNT(9),LASTS(9),Y(20000),XMISS(512)
      N=LEVEL(1)
      DO 150 I=2,K
 150  N=N*LEVEL(I)
      LASTS(1)=LEVEL(1)
      DO 178 I=2,K
 178  LASTS(I)=LEVEL(I)+1
      NN=1
      LL=(2**K)-1
      MSTEP(1)=1
      DO 180 I=2,K
 180  MSTEP(I)=MSTEP(I-1)*2
      DO 185 I=1,LL
      XDF(I)=16.E60
      XMISS(I)=0.0
 185  SUMSQ(I)=0.0
      DO 190 I=1,K
 190  KOUNT(I)=0
 200  L=0
      DO 260 I=1,K
      IF(KOUNT(I)-LASTS(I)) 210, 250, 210
 210  IF(L) 220, 220, 240
 220  KOUNT(I)=KOUNT(I)+1
      IF(KOUNT(I)-LEVEL(I)) 230, 230, 250
 230  L=L+MSTEP(I)
```

150

```
      GO TO 260
240 IF(KOUNT(I)-LEVEL(I)) 230, 260, 230
250 KOUNT(I)=0
260 CONTINUE
      IF(L) 285, 285, 270
270 XMISS( L)=XMISS( L)+AMIN1(X(NN),Y(NN))
      IF(NN.GT.20000)PRINT *,X(NN),Y(NN),NN,L
       IF(X(NN).LT.0..OR.Y(NN).LT.0.)PRINT*,L,NN,X(NN),Y(NN)
      IF(X(NN)-Y(NN) .EQ.0) GO TO 275
      XX=ABS(SQRT(X(NN))-SQRT(Y(NN)))
                                    XDF(L)=AMIN1(XDF(L),XX)
      SUMSQ(L)=SUMSQ(L)+XX
275 NN=NN+1
      GO TO 200
285 FN=N
       IF(NN.GT.20000)PRINT*,L,X(NN),Y(NN),NN
      GMEAN=X(NN)+Y(NN)
      DO 310 I=2,K
310 MSTEP(I)=0
      NN=0
      MSTEP(1)=1
320 ND1=1
      ND2=1
      DO 340 I=1,K
      IF(MSTEP(I)) 330, 340, 330
330 ND1=ND1*LEVEL(I)
340 CONTINUE
      ND2=ND1
      FN1=ND1
      FN2=ND2
      NN=NN+1
      SUMSQ(NN)=SUMSQ(NN)/(SQRT(FN/FN1))
      XDF(NN)=XDF(NN)/(SQRT(FN/FN1))
      SMEAN(NN)=SUMSQ(NN)/FN2
      IF(NN-LL) 345, 370, 370
345 DO 360 I=1,K
      IF(MSTEP(I)) 347, 350, 347
347 MSTEP(I)=0
      GO TO 360
350 MSTEP(I)=1
      GO TO 320
360 CONTINUE
370 RETURN
      END
      SUBROUTINE AVCAL (K,LEVEL,X,L,ISTEP,LASTS,Y)
      DIMENSION LEVEL(9),ISTEP(9),LASTS(9)
      DIMENSION X(20000),Y(20000)
      LASTS(1)=L+1
      DO 645 I=2,K
645 LASTS(I)=LASTS(I-1)+ISTEP(I)
650 DO 175 I=1,K
      L=1
      ISUMX=0.0
      ISUMY=0.0
      NN=LEVEL(I)
      INCRE=ISTEP(I)
      LAST=LASTS(I)
155 DO 660 J=1,NN
      ISUMX=ISUMX+X(L)
      ISUMY=ISUMY+Y(L)
```

151

```
      L=L+INCRE
660   CONTINUE
      X(L)=ISUMX
      Y(L)=ISUMY
      ISUMX=0.0
      ISUMY=0.0
      IF(L-LAST) 167, 175, 175
167   IF(L-LAST+INCRE) 168, 168, 670
168   L=L+INCRE
      GO TO 155
670   L=L+INCRE+1-LAST
      GO TO 155
175   CONTINUE
      RETURN
      END
      SUBROUTINE AVDAT (K,LEVEL,N,X,L,ISTEP,KOUNT,Y)
      DIMENSION LEVEL(9),ISTEP(9),KOUNT(9)
      DIMENSION X(20000),Y(20000)
      M=LEVEL(1)+1
      DO 105 I=2,K
105   M=M*(LEVEL(I)+1)
      N1=M+1
      N2=N+1
      DO 107 I=1,N
      N1=N1-1
      N2=N2-1
      Y(N1)=Y(N2)
107   X(N1)=X(N2)
      ISTEP(1)=1
      DO 110 I=2,K
110   ISTEP(I)=ISTEP(I-1)*(LEVEL(I-1)+1)
      DO 115 I=1,K
115   KOUNT(I)=1
      N1=N1-1
      DO 135 I=1,N
      L=KOUNT(1)
      DO 120 J=2,K
120   L=L+ISTEP(J)*(KOUNT(J)-1)
      N1=N1+1
      X(L)=X(N1)
      Y(L)=Y(N1)
      DO 130 J=1,K
      IF(KOUNT(J)-LEVEL(J)) 124, 125, 124
124   KOUNT(J)=KOUNT(J)+1
      GO TO 135
125   KOUNT(J)=1
130   CONTINUE
135   CONTINUE
      RETURN
      END
```

6.6 PROG. LACHIN

Description: This program, written by Professor John M. Lachin, selects "good" subsets of variables on the basis of the change in the degree of association between the grouping variable, having two levels, and the remaining variables when an additional variable is added to the predictor set. At each step the program computes a "full χ^2" and a "partial χ^2" for each of the possible predictors. Next, that the variable with the most significant partial χ^2 is added to the predictor subset at each step, provided that it is significant beyond a specified level (MPA). If at a step, none of the partial χ^2 values exceed the MPA, the process terminates. In addition, before each step the program checks the partial χ^2 values for those variables having been selected and if any fails to yield a partial χ^2 that is significant beyond a specified level (MPD), then that variable is deleted from the predictor set at that step. After the selection process terminates the program computes the allocation rule by the usual likelihood-ratio criterion. The program begins this stage by computing the estimates of the $P\{\mathbf{x}|G_i\}$, $i=1$, 2 for each $\mathbf{x}$. The indices U are constructed by treating the values of the k selected or specified variables as imaginary subscripts of a k-dimensional space. From these k subscripts a single subscript is computed by successive subscript reduction. These single subscripts are then ranked and the rank of each subscript is used as the value of U. The likelihood ratio for each U is calculated and then also ranked. Input to the program is the user's original data matrix coded in 1, 2,...,k form (corresponding to the levels of the variables) along with some additional control information. Note that the grouping variable must also be coded in 1, 2 form, denoting the first or second group.

Card Preparation: The following cards are used to run LACHIN:

1. *Problem Card*

Columns	Contents
1–7	"PROBLEM"
8	Blank
9–11	Number of input variables $\leqslant 50$
12–13	Input logical unit 5 if data are to be read from cards; logical unit 1 cannot be so used since it is reserved for LACHIN
14	Number of variable format cards to follow $\leqslant 3$
15–16	Number of case deletion criteria ($\leqslant 18$) to be specified on a DELETION card; if zero, a DELETION card should not be used
17–80	Any alphanumeric problem description

2. *Maximum Variable Values Card(s): All values for all variables must be
   > 0.*

Columns	Contents
1–8	"MAXVALUE"
9–10	The maximum value of variable 1
11–12	The maximum value of variable 2
13–14	The maximum value of variable 3
. . . . .	
79–80	The maximum value of variable 36
	Continue onto an additional card if necessary. The additional card should have "MAXVALUE" in columns 1–8

3. *Variable Format Card(s).* Using all 80 columns of the number of cards
   specified in column 14 of the problem card, specify the location of each
   variable in I format.

4. *Deletion Card.* This card should be used *only* if a nonzero value is
   specified in columns 15–16 of the PROBLEM card.

Columns	Contents
1–8	"DELETION"
9–10, 11–12	The first deletion criteria
13–14, 15–16	The second deletion criteria
. . . .	
77–78, 79–80	The eighteenth deletion criteria

Each deletion criterion is allocated four spaces. The first two spaces contain
the variable number (V), and the second two spaces contain the value of
response (R). Any case with the value R for the variable V will then be
deleted from the decision function analysis in any problem in which the
variable V is specified as a predictor on the PREDCTRS card.

5. *Subproblem Card*

Columns	Contents
1–8	"SUBPRBLM"
9–10	The variable number of the criterion variable
11–12	The number of predictor variables $\leqslant 50$
13	Specify whether or not the stepwise procedure is to be applied. 0 No 1 Yes

Columns	Contents
14–17	If column 13 contains a "1," specify the probability level to be used for the addition of a variable (MPA) in F4.3 format (i.e., as ".xxx").
18–21	If column 13 contains a "1," specify the probability level to be used for the deletion of a variable already selected (MPD) (MPD ⩾ MPA) in F4.3 format (i.e., as ".xxx").
22–80	Any alphanumeric subproblem description.

6. *Subproblem Group Parameter Specification Card*

Columns	Contents
1–8	"PARAMETR"
9–16	The name of the first group (criterion variable = 1)
17–24	The name of the second group (criterion variable = 2)
25–34	The *a priori* probability of membership in the first group in F10.8 format (keypunch decimal)
35–44	The *a priori* probability of membership in the second group in F10.8 format (keypunch decimal)

7. *Subproblem Predictor Specification Card(s)*

Columns	Contents
1–8	"PREDCTRS"
9–10	The variable number of the first predictor
11–12	The variable number of the second predictor
. . . . .	
79–80	The variable number of the 36th predictor
	Continue onto an additional card if necessary; the additional card should have "PREDCTRS" in columns 1–8

8. *New Problem Card.* To be specified if an additional problem is to be initiated, followed immediately by a new PROBLEM card

Columns	Contents
1–8	"NEWPRBLM"

9. *Finish Card*

Columns	Contents
1–6	"FINISH"
7–8	Blank

Output: Along with a printout of all input parameters the program provides the following results.

1. A listing at each step of both full and partial x^2 for all variables, associated probabilities, and that variable to be included or deleted.
2. The results of the likelihood-ratio analysis; starting with the smallest likelihood, the values of U, the likelihood $(\hat{P}(U|G_1)/\hat{P}(U|G_2))$, the conditional probabilities $(\hat{P}(U|G_i),\ i=1,\ 2)$ and the posterior probabilities $(\hat{P}(G_i|U),\ i=1,\ 2)$ are listed, and also listed are the cumulative $\hat{P}(U|G_1)$ and $\hat{P}(U|G_2)$ (indicating the probability of correctly classifying members of the two respective groups), and the percent correctly classified in the entire sample are listed.
3. Following the likelihood-ratio analysis, a conversion table is listed that gives the values of U with the corresponding likelihood and group assignment and the response patterns.

Uses and Limitations: The program is dimensioned to handle 50 variables. However, it is restricted to handle at most 500 nonempty response patterns. All other program restrictions have been noted in the description of each parameter card.

```
PROG, LACHIN
      COMMON NVAR,JND(50),KND,IVAL(50),MAXVAL(50),IND,INDICS(50,500)
      INTEGER*2 IND,INDICS
      INTEGER*2 MAXVAL,IVAL,JND,KND
      COMMON /SORT/ ITAB(1000),JTAB(1000)
      REAL*4 ITAB,JTAB
      COMMON /THTA/ THETA,FR(2),VPC(2),TN(2)
      REAL*8 GPLABL(2),FMT(30),BLANK/'          '/,IDA,PRO
      REAL*8 TCRD,PRB/'PROBLEM '/,MVL/'MAXVALUE'/,SUBP/'SUBPRBLM'/,
     AFIN/'FINISH  '/,PRD/'PREDCTRS'/,PRM/'PARAMETR'/
      REAL*8 NWP/'NEWPRBLM'/,DEL/'DELETION'/,NWC/'NEWCHISQ'/
      DIMENSION VPT(2),PRIORI(2),PCT(2)
      INTEGER*2 FRQ(50,2,500),IVARS(50),IV(50),MAXV(50)
      INTEGER*2 KARRAY(500),NSV,ISV(20),ISVV(20)
      LOGICAL RESEL,HIT,NODATA
    1 RESEL=.FALSE.
      TCRD=BLANK
      TN(1)=0.0
      TN(2)=0.0
      READ (5,500) TCRD,NV,IN,NFMT,NSV,(FMT(I),I=1,8)
  500 FORMAT (A8,I3,I2,I1,I2,8A8)
      PRO=PRB
      IF(TCRD.NE.PRO) GO TO 1550
      WRITE (6,550) (FMT(I),I=1,8),NV,IN,NFMT,NSV
  550 FORMAT (1H1,'STATISTICAL DECISION FUNCTION FOR TWO GROUPS',//,
     / ' JOHN M. LACHIN' ,/,
     A ' THE BIOSTATISTICS CENTER, DEPARTMENT OF STATISTICS, THE GEORGE
     BWASHINGTON UNIVERSITY',///,
     C ' PROBLEM DESCRIPTION.....',8A8,//,
     D ' NUMBER OF VARIABLES.....',I5,//,
     E ' INPUT LOGICAL UNIT......',I5,//,
     F ' NO. VARIABLE FORMAT CARDS',I4,//,
     G ' NO. CASE DELETION CRITERIA',I3)
      IF(NV-50) 300,300,301
  301 WRITE (6,302)
  302 FORMAT (1H0,'THE NUMBER OF VARIABLES IS GREATER THAN 50. JOB ABORT
     AED.')
      GO TO 1606
  300 IF(IN-1) 304,303,304
  303 WRITE (6,350)
  350 FORMAT (1H0,'LOGICAL UNIT 1 CANNOT BE USED FOR DATA INPUT SINCE IT
     A IS RESERVED FOR USE BY NEYMAN. JOB ABORTED.')
      GO TO 1606
  304 IF(NFMT-3) 305,305,306
  306 WRITE (6,354)
  354 FORMAT (1H0,'THE NUMBER OF VARIABLE FORMAT CARDS IS GREATER THAN 3
     A. JOB ABORTED.')
      GO TO 1606
  305 IF(NSV-18) 255,255,256
  256 WRITE (6,653)
  653 FORMAT (1H0,'NO MORE THAN 18 CASE DELETION CRITERIA CAN BE SPECIFI
     AED. JOB ABORTED.')
      GO TO 1606
  255 TCRD = BLANK
      PRO=MVL
      IF(NV-36) 560,561,560
  561 READ (5,562) TCRD,(MAXV(I),I=1,NV)
  562 FORMAT (A8,36I2)
      GO TO 563
  560 READ (5,537) TCRD,(MAXV(I),I=1,NV)
```

157

```
    537 FORMAT (A8,36I2,(/,8X,36I2))
    563 IF(TCRD.NE.PRO) GO TO 1550
        WRITE (6,551) (MAXV(I),I=1,NV)
    551 FORMAT (1H0,'MAXIMUM VARIABLE VALUES.',35I3,2(/,25X,35I3))
          NFMT=NFMT*10
          DO 12 I=1,30
     12 FMT(I)=BLANK
        READ (5,501) (FMT(I),I=1,NFMT)
    501 FORMAT (10A8)
        WRITE (6,552) (FMT(I),I=1,NFMT)
    552 FORMAT (1H0,'VARIABLE FORMAT.........',10A8,2(/,25X,10A8))
     74 CONTINUE
        IF(NSV) 253,253,254
    254 PRO=DEL
        READ (5,522) TCRD,(ISV(I),ISVV(I),I=1,NSV)
    522 FORMAT (A8,36I2)
        IF(TCRD.NE.PRO) GO TO 1550
        WRITE(6,523) NSV,(ISV(I),ISVV(I),I=1,NSV)
    523 FORMAT (1H0,10X,'IF ANY OBSERVATION CONTAINS ONE OF THE FOLLOWING
       A',I3,' RESPONSES (R) ON THE STATED VARIABLES (V - R)',/,11X,
       C 'IT WILL BE DELETED FROM THIS PROBLEM.',/,11X, 7('(',I3,' - ',
       D I3,')',5X),/,11X,7('(',I3,' - ',I3,')',5X))
    253 NSEL=0
     53 TCRD=BLANK
        PRO=SUBP
        READ (5,507,END=54)   TCRD,NBS,NVR,ISTP,PADD,PDEL,(ITAB(I),I=1,15)
    507 FORMAT (A8,2I2,I1,2F4.3,14A4,A3)
        IF(TCRD.EQ.NWP) GO TO 1
        IF(TCRD.NE.PRO) GO TO 1550
        NSEL=NSEL+1
        WRITE (6,510) NSEL,NBS,NVR
    510 FORMAT (1H1,'SUBPROBLEM NO. ',I2,', USING VARIABLE NO. ',I3,
       A ' AS THE CRITERION VARIABLE AND ',I3,' PREDICTORS.')
        WRITE (6,570) (ITAB(I),I=1,15)
    570 FORMAT (1H0,10X,'SUBPROBLEM DESCRIPTION - ',14A4,A3)
        IF(ISTP) 250,250,251
    250 WRITE (6,650)
    650 FORMAT (1H0,10X,'THE STEPWISE OPTION WILL NOT BE EXERCISED. THE DE
       ACISION FUNCTION WILL BE BASED ON ALL THE SPECIFIED PREDICTORS.')
        GO TO 252
    251 WRITE (6,651) PADD,PDEL
    651 FORMAT (1H0,10X,'THE STEPWISE PROCEDURE WILL BE APPLIED WITH MINIM
       AUM PROBABILITIES',//,16X,F5.3,' FOR THE ADDITION OF A VARIABLE TO
       BTHE PREDICTOR SET, AND',//,16X,F5.3, ' FOR THE DELETION OF A VARIA
       CBLE FROM THE PREDICTOR SET.')
    252 IF(NVR-50) 576,576,577
    577 WRITE (6,578)
    578 FORMAT (1H0,'THE NUMBER OF PREDICTORS IS GREATER THAN 50. JOB ABOR
        ATED.')
        GO TO 1606
    576 CONTINUE
        IF(MAXV(NBS).EQ.2) GO TO 573
        WRITE (6,574) NBS
    574 FORMAT (1H0,'THE MAXIMUM VALUE SPECIFIED FOR THE CRITERION VARIABL
       AE ',I3,' IS NOT 2. JOB ABORTED.')
        GO TO 1606
    573 IF(ISTP.LE.0) GO TO 257
        IF(PDEL.GE.PADD) GO TO 257
        WRITE (6,652)
    652 FORMAT (1H0,'THE MINIMUM PROBABILITY FOR DELETION IS LESS THAN THA
```

```
      AT FOR ADDITION. JOB ABORTED.')
      GO TO 1606
  257 CONTINUE
      TCRD=BLANK
      PRO=PRM
      READ (5,511) TCRD,(GPLABL(I),I=1,2),(PRIORI(I),I=1,2)
  511 FORMAT (3A8,2F10.8)
      IF(TCRD.NE.PRO) GO TO 1550
      TCRD=BLANK
      PRO=PRD
      IF(NVR-36) 564,565,564
  565 READ (5,562) TCRD,(IVARS(I),I=1,NVR)
      GO TO 566
  564 READ (5,537) TCRD,(IVARS(I),I=1,NVR)
  566 IF(TCRD.NE.PRO) GO TO 1550
      WRITE (6,572) (IVARS(I),I=1,NVR)
  572 FORMAT (1H0,10X,'THE PREDICTORS ARE VARIABLES ',30I3,
     A 2(/,40X,30I3))
      DO 25 I=1,2
   25 WRITE (6,512) GPLABL(I),I,PRIORI(I)
  512 FORMAT (1H0,10X,'THE A PRIORI PROBABILITY OF MEMBERSHIP IN THE ',
     A A8,' GROUP, P(I=',I1,'), IS ',F6.2)
      WRITE (6,517) GPLABL(1)
  517 FORMAT (1H0,10X,'THE NULL HYPOTHESIS IS EQUATED WITH MEMBERSHIP IN
     A THE ',A8,' GROUP.')
      WRITE (6,518)
  518 FORMAT (1H0,10X,'IF C(I) IS THE RELATIVE COST OF MISCLASSIFYING A
     A CASE BELONGING TO GROUP I, THEN',
     /                          //,16X,'THE THRESHOLD VALUE OF LAMBDA =
     B ( P(I=2) * C(I=2) ) / ( P(I=1) * C(I=1) )')
      THRESH=PRIORI(2)/PRIORI(1)
      WRITE (6,519) THRESH
  519 FORMAT (1H0,45X,'= ',F15.5,' * ( C(I=2) / C(I=1) )')
   84 CONTINUE
      NPDS=NVR
      IF(ISTP.GT.0) GO TO 105
      NVAR=NVR
      NVR=1
      DO 152 I=1,NVAR
  152 MAXVAL(I) = MAXV(IVARS(I))
      GO TO 101
  105 SCHISQ=0.0
        SDF=0.0
      NIN=0
      NSTP=0
C
C
      INUM=0
  100 NSTP=NSTP+1
      WRITE (6,600) NSTP,NIN,SCHISQ,SDF
  600    FORMAT (1H1,'STEP NO.',I3,/,10X,'THE CHI-SQUARE FOR THE',
     C  I3,' VARIABLES SELECTED AS PREDICTORS IS',F15.8,
     C   ' ON ',F5.0,' DF.',/)
      NIN=NIN+1
        SDF=SIGN(SDF,-1.0)
        SCHISQ=SIGN(SCHISQ,-1.0)
        CHISQM=0.0
      PCHIM=1.0
      WRITE (6,655)
  655 FORMAT (1H0,' VARIABLES',4X,'VARIABLES',9X,'FULL CHI-SQUARE',
```

```
      A 22X,'PARTIAL CHI-SQUARE',/,4X,'ADDED',6X,'NOT ADDED',10X,
      B 'CHI-SQ.',5X,'DF',24X,'CHI-SQ.',5X,'DF',
      C 6X,'PROB.',/)
610 FORMAT (1H ,I7,61X,F10.4,I7,E13.3)
601 FORMAT (1H ,I7,23X,F10.4,I7,21X,F10.4,I7,E13.3)
603 FORMAT (1H ,17X,I3,10X,F10.4,I7,21X,F10.4,I7,E13.3)
      NVMR=NIN-2
101 DO 15 IP=1,NVR
      JND(IP)=0
      DO 15 I=1,500
      INDICS(IP,I)=0
      DO 15 IG=1,2
 15 FRQ(IP,IG,I)=0
  5 IF(RESEL) GO TO 50
      READ (IN,FMT,END=6)        (IV(I),I=1,NV)
      IF(NSV.EQ.0) GO TO 82
      DO 83 I=1,NSV
      IF(IV(ISV(I)).EQ.ISVV(I)) GO TO 5
 83 CONTINUE
 82 CONTINUE
      NODATA=.FALSE.
      HIT=.FALSE.
      DO 7 I=1,NPDS
      J=IVARS(I)
      IF(IV(J).LE.0) NODATA=.TRUE.
      IF(IV(J).GT.MAXV(J)) HIT=.TRUE.
  7 CONTINUE
      IG=IV(NBS)
      IF(IG.LE.0) NODATA=.TRUE.
      IF(IG.GT.MAXV(NBS)) HIT=.TRUE.
      INUM=INUM+1
      IF(NODATA) GO TO 70
      IF(HIT) GO TO 270
      GO TO 71
270 WRITE(6,670) INUM
670 FORMAT (' CASE SEQUENCE NUMBER ',I5,' WILL BE DELETED DUE TO DATA
     AOUT OF RANGE (VALUE .GT. MAXIMUM) FOR AT LEAST',/,5X,
     B 'ONE VARIABLE USED IN THIS SUBPROBLEM.')
      GO TO 5
 70 WRITE (6,521) INUM
521 FORMAT (' CASE SEQUENCE NUMBER ',I5 ,' WILL BE DELETED DUE TO MISS
     AING DATA (.LE.0) FOR AT LEAST ONE VARIABLE USED IN THIS SUBPROBLEM
     B.')
      IF (HIT) GO TO 270
      GO TO 5
 71 TN(IG)=TN(IG)+1.0
      WRITE (1) (IV(I),I=1,NV)
      GO TO 51
 50 READ (1,END=60) (IV(I),I=1,NV)
      IG=IV(NBS)
 51 IF(ISTP.GT.0) GO TO 111
      DO 17 I=1,NPDS
      J=IVARS(I)
 17 IVAL(I)=IV(J)
      CALL INDEX
      CALL CHECK (1)
      FRQ(1,IG,KND)=FRQ(1,IG,KND)+1
      GO TO 5
111 NVAR=NVMR
      DO 700 IP=1,NVR
```

```
      IF(IP-NIN) 701,702,703
  701 IF(NSTP-2) 707,707,704
  707 MAXVAL(1)=MAXV(IVARS(1))
      IVAL(1)=IV(IVARS(1))
      GO TO 700
  704 I=IVARS(1)
      DO 705 J=1,NVAR
      IVARS(J)=IVARS(J+1)
      IVAL(J)=IV(IVARS(J))
  705 MAXVAL(J)=MAXV(IVARS(J))
      IVARS(NVAR+1)=I
      MAXVAL(NVAR+1)=MAXV(I)
      IVAL(NVAR+1)=IV(I)
      IF(NIN.LT.NSTP) GO TO 706
      IF(IP.EQ.NIN-1) GO TO 700
      GO TO 706
  702 NVAR=NIN
      JVNV=IVARS(IP)
  703 MAXVAL(NVAR)=MAXV(IVARS(IP))
      IVAL(NVAR)=IV(IVARS(IP))
      IVARS(NVAR)=IVARS(IP)
  706 CALL INDEX
      I=IP
      CALL CHECK(I)
  757 FRQ(IP,IG,KND)=FRQ(IP,IG,KND)+1
  700 CONTINUE
      IVARS(NVAR)=JVNV
      GO TO 5
    6 ENDFILE 1
      RESEL =.TRUE.
      TOTN=TN(1)+TN(2)
      PCT(1)=TN(1)/TOTN
      PCT(2)=TN(2)/TOTN
   60 REWIND 1
      IF(ISTP.EQ.0) GO TO 112
      NVAR=NIN-2
      DO 750 IP=1,NVR
      IF(IP-NIN) 751,752,753
  751 IF(NSTP-2) 754,754,755
  754 MAXVAL(1)=MAXV(IVARS(1))
      IDDF=DFM2
      WRITE (6,610) IVARS(IP),CHSQM2,IDDF,PCHM2
      GO TO 750
  755 I=IVARS(1)
      DO 756 J=1,NVAR
      IVARS(J)=IVARS(J+1)
  756 MAXVAL(J)=MAXV(IVARS(J))
      IVARS(NVAR+1)=I
      NVTST=NVAR+1
      IF(NIN.LT.NSTP) GO TO 758
      IF(IP.NE.NIN-1) GO TO 758
      CHISQ=CHSQM3
      IDF=DFM3
      PCHI=PCHM3
      DCHISQ=CHSQM2
      IDDF=DFM2
      DPCHI=PCHM2
      GO TO 120
  752 NVAR=NIN
      SCHISQ=SIGN(SCHISQ,+1.0)
```

```
      SDF=SIGN(SDF,+1.0)
      JVNV=IVARS(IP)
  753 MAXVAL(NVAR)=MAXV(IVARS(IP))
      IVARS(NVAR)=IVARS(IP)
  758 CHISQ=0.0
      MAX=JND(IP)
        DO 104 I=1,MAX
      FR(1)=FRQ(IP,1,I)
      FR(2)=FRQ(IP,2,I)
      TFR=FR(1)+FR(2)

      DO 81 K=1,2
    COMPUTE THE SIMPLE GOODNESS OF FIT CHI-SQUARE
      TEXP=PCT(K)*TFR
        IF(TEXP.EQ.0.0) GO TO 81
      CHISQ=CHISQ+(((FR(K)-TEXP)**2)/TEXP)
   81 CONTINUE


  104  CONTINUE
      DF=0.0

    FIND THE SIZE OF THE JOINT DISTRIBUTION SPACE

      IDF=MAXVAL(1)
      IF(NVAR.EQ.1) GO TO 80
      DO 75 I=2,NVAR
      ID=MAXVAL(I)
   75 IDF=IDF*ID
   80 CONTINUE
C
C     DF AS THE NUMBER OF NON-NULL SAMPLE POINTS MINUS ONE
C  PLUS ONE FOR THE SET OF REMAINING NULL SAMPLE POINTS
C   IF SUCH A SET EXISTS.
C
      IF(MAX-IDF) 76,77,78
   76 IDF=MAX
      GO TO 79
   77 IDF=MAX-1
      GO TO 79
   78 IDF=0
   79 DF=IDF
      D=0.0
      IER=0
      IDF=DF
      DCHISQ = SIGN(CHISQ,SCHISQ) - SCHISQ
      DDF= SIGN(DF,SDF) - SDF
      CALL CDTR( DCHISQ,DDF,P,D,IER)
      DPCHI=1.0-P
      IDDF = DDF
      IF (IP-NIN) 120,121,121
  120 WRITE (6,601) IVARS(NVTST),CHISQ,IDF,DCHISQ,IDDF,DPCHI
      IF (DPCHI.LE.PDEL) GO TO 750
      SCHISQ=CHISQ
      SDF=IDF
      NIN=NIN-2
      WRITE (6,602) IVARS(NVTST)
  602 FORMAT (1H0,'VARIABLE ',I5,' WILL BE DELETED. THIS STEP IS COMPLET
     AED.')
      GO TO 100
```

```
121 WRITE (6,603) IVARS(IP),CHISQ,IDF,DCHISQ,IDDF,DPCHI
    IF (DPCHI-PCHIM) 123,124,122
124 TEMP=DCHISQ/DDF
    TEXP=CHSQM2/DFM2
    IF(TEMP.LE.TEXP) GO TO 122
123 PCHIM=DPCHI
    CHISQM=CHISQ
    DFM=DF
    IVM=IP
    CHSQM2=DCHISQ
    DFM2=DDF
    PCHM2=DPCHI
122 CONTINUE
750 CONTINUE
    IVARS(NVAR) = JVNV
    IF (PCHIM-PADD) 126,126,127
126 WRITE (6,605) IVARS(IVM)
605 FORMAT (1H0,'VARIABLE ',I5,' WILL BE ADDED TO THE SET OF PREDICTOR
   AS AT THIS STEP.')
    CHSQM3=SCHISQ
    DFM3=SDF
    SCHISQ=CHISQM
    SDF=DFM
    I=IVARS(IVM)
    K=IVM-NIN
    DO 130 J=1,K
130 IVARS(IVM-J+1) = IVARS(IVM-J)
    IVARS(NIN) = I
    GO TO 100
127 ISTP=0
    WRITE (6,615)
615 FORMAT (1H0,'NO ADDITIONAL VARIABLES WILL BE ADDED. THE STEPWISE P
   AROCEDURE IS COMPLETED.')
    NVAR=NVAR-1
    NVR=1
    IF(NVAR.GT.0) GO TO 101
    WRITE (6,606)
    GO TO 1606
112 MAX=JND(1)
103 INC=-1
606 FORMAT ('0THE STEPWISE PROCEDURE DID NOT SELECT ANY PREDICTORS. JOB
   /B ABORTED. CHANGE PADD, PDEL AND RESUBMIT.')
    JNC=0
    HIT=.FALSE.
     DO 16 I=1,1000
     ITAB(I)=0.0
 16  JTAB(I)=0.0
    DO 8 I=1,MAX
    INC=INC+2
    JNC=JNC+2
    DO 20 K=1,2
    FR(K)=FRQ(1,K,I)
 20 CONTINUE
    CALL COMPT
 10 ITAB(INC)=THETA
  8 ITAB(JNC)=I
    NSB2=JNC
    CALL NSORT (JNC)
    J=NSB2+2
    DO 30 I=1,MAX
```

```
        J=J-2
        K=J-1
        KND=ITAB(J)
        KARRAY(I)=KND
  30 ITAB(K)=INDICS(1,KND)
        CALL NSORT(NSB2)
        J=NSB2+2
        DO 31 I=1,MAX
        J=J-2
        KND=ITAB(J)
  31 JTAB(KND)=I
        WRITE (6,504)
 504 FORMAT (1H1,'THE LIKELIHOOD RATIO (LAMBDA) ANALYSIS' )
        WRITE (6,505) (GPLABL(I),I=1,2)
 505 FORMAT (1H0,39X,A8,32X,A8,23X,'PERCENT CORRECT',/,5X,1HU,
     A 8X,'LAMBDA',7X,'FREQ.   P(U!I)      P(I!U)      ALPHA',
     B            7X,'FREQ.   P(U!I)      P(I!U)     1-BETA',
     C 8X,'SAMPLE      POPULATION')
 506 FORMAT (1H ,I5,5X,F10.5,2(F10.0,3F10.5),F15.5,F12.5)
        J=NSB2+2
        ALPHA=0.0
        BETA=0.0
        TC=TN(1)
        TP=PRIORI(1)
        DO 11 I=1,MAX
        J=J-2
        L=J-1
        KND=KARRAY(I)
        JTP=JTAB(KND)
        DO 210 K=1,2
        FR(K)=FRQ(1,K,KND)
 210 CONTINUE
        CALL COMPT
        DO 21 K=1,2
        VPT(K)=0.0
        IF(FR(K).EQ.0.0) GO TO 21
        VPT(K)=VPC(K)*PRIORI(K)
  21 CONTINUE
        IF(THETA.LT.0.0) THETA=1.0E-10
        IF(THETA.GT.100.0) THETA=1.0E10
        ALPHA=ALPHA+VPC(1)
        BETA=BETA+VPC(2)
        TC=TC+FR(2)-FR(1)
        VTC=(TC/TOTN)*100.0
        TP=TP-VPT(1)+VPT(2)
        VTP=TP*100.0
        TEMP=VPT(1)+VPT(2)
        TFR=FR(1)+FR(2)
        DO 40 K=1,2
        IF(VPT(K).EQ.0.0) GO TO 40
        VPT(K)=VPT(K)/TEMP
  40 CONTINUE
        K=MAX+JTP
        JTAB(K)=THETA
        IF(THETA.LT.THRESH) GO TO 11
        IF(HIT) GO TO 11
        HIT=.TRUE.
        WRITE (6,520)
 520 FORMAT (1H0,44(1H*),' CUTOFF - ASSUMING THAT C(I=2) = C(I=1) ',
     A 47(1H*),/)
```

164

```
   11 WRITE (6,506) JTP,THETA,FR(1),VPC(1),VPT(1),ALPHA,FR(2),VPC(2),
     A VPT(2),BETA,VTC,VTP
        TC=TN(1)+TN(2)
        DO 26 K=1,2
   26 VPC(K)=TN(K)/TC
        WRITE (6,516) (TN(K),VPC(K),K=1,2)
  516 FORMAT (1H ,'TOTAL',15X,2(F10.0,F10.5,20X))
        WRITE (6,513)
  513 FORMAT (1H0,'P(U!I) = THE SAMPLE ESTIMATE OF THE PROBABILITY THAT
     AA CASE DRAWN RANDOMLY FROM GROUP I WILL HAVE RESPONSE PATTERN U.')
        WRITE (6,514)
  514 FORMAT (1H0,'P(I!U) = THE PROBABILITY THAT A CASE DRAWN RANDOMLY F
     AROM THE POPULATION, AND HAVING RESPONSE PATTERN U,',/,
     B 10X,'IS A MEMBER OF THE GROUP I.')
        WRITE (6,599)
  599 FORMAT (1H1)
        WRITE (6,502)
  502 FORMAT (1H0,'THE CONVERSION TABLE SHOWING THE INDEX (U)',
     A                                           ' ASSOCIATED WIT
     AH EACH RESPONSE PATTERN ENCOUNTERED AND THE DECISION FUNCTION.')
        WRITE (6,503) (IVARS(I),I=1,NVAR)
  503 FORMAT (1H0,18X,'GROUP',6X,'VARIABLES',/,4X,1HU,3X,'LAMBDA',
     A 4X,'DENOTED',2X,35I3,/,(27X,35I3))
        J=NSB2+2
        DO 13 I=1,MAX
        J=J-2
        KND=ITAB(J)
        IND=INDICS(1,KND)
        THETA=JTAB(MAX+I)
        CALL DINDEX
        FMT(1)=GPLABL(2)
        IF(THETA.GE.THRESH) FMT(1)=GPLABL(1)
   13 WRITE (6,508) I,THETA,FMT(1),(IVAL(K),K=1,NVAR)
  508 FORMAT (1H ,I4,3X,F7.2,2X,A8,2X,35I3,/,(27X,35I3))
        GO TO 53
   54 CONTINUE
 1550 IF(TCRD.EQ.FIN) GO TO 1552
        WRITE (6,1551) TCRD,PRO
 1551 FORMAT (1H0,'THE FIRST 8 COLUMNS OF THE LAST CARD READ CONTAIN ',
     A A8,' WHILE A ',A8,' CARD WAS EXPECTED. JOB ABORTED.')
        GO TO 1606
 1552 WRITE (6,1553)
 1553 FORMAT (1H0,'FINISH CARD ENCOUNTERED. JOB COMPLETED.')
 1606 CONTINUE
        STOP
        END
        SUBROUTINE NSORT (NSB2)
        COMMON /SORT/ ITAB(1000),JTAB(1000)
        REAL*4 ITAB,JTAB,ITEMP
        J1=1
        J2=NSB2-1
C     HIGH SPEED SORT BY REPEATED MERGES OF STRINGS OF DOUBLING LENGTH  DISC4240
C     SORT THE SUBJECTS BY DISCRIMINANT SCORES                          DISC4250
C     TABLE TO BE SORTED GOES FROM J1 TO J2 BY PAIRS                     DISC4260
C     START MERGING STRINGS OF LENGTH 1                                  DISC4270
 1210 IDBL=2                                                             DISC4280
C     REENTER HERE FOR EACH VALUE OF IDBL=2*(1,2,4,8,...) .LE. NSB2      DISC4290
 1220 J2B=J1-2
        K=-1                                                             DISC4310
C     REENTER HERE FOR EACH PAIR OF STRINGS TO BE MERGED                DISC4320
```

165

```
C     DETERMINE UPPER AND LOWER INDICES OF THESE STRINGS            DISC4330
 1230 J1A=J2B+2                                                      DISC4340
      J1B=J1A+IDBL                                                   DISC4350
      J2A=J1B-2                                                      DISC4360
      J2B=J2A+IDBL                                                   DISC4370
C     SPECIAL END TESTS FOR ONE OR BOTH STRINGS SHORTER THAN IDBL   DISC4380
C     TEST IF THERE ARE ANY ITEMS IN THE SECOND STRING              DISC4390
      IF(J2-J2A) 1240,1240,1250                                     DISC4400
 1240 J2A=J2                                                         DISC4410
      GO TO 1330                                                     DISC4420
C     TEST IF LENGTH OF SECOND STRING IS LESS THAN IDBL             DISC4430
 1250 J2B=MINO(J2B,J2)                                               DISC4440
C     ENTER MERGE OF (ITAB(J1A),...,ITAB(J2A))                      DISC4450
C     WITH (ITAB(J1B),...,ITAB(J2B)). MERGED STRING INTO JTAB.      DISC4460
      I=J1B                                                          DISC4470
      GO TO 1360                                                     DISC4480
C     OUTPUT FROM FIRST STRING UNTIL ITEMP NOT EXCEEDED             DISC4490
 1260 DO 1280 I=J1A,J2A,2                                            DISC4500
      IF(ITAB(I)-ITEMP) 1300,1270,1270                              DISC4510
 1270 K=K+2                                                          DISC4520
      JTAB(K)=ITAB(I)                                                DISC4530
 1280 JTAB(K+1)=ITAB(I+1)                                            DISC4540
C     OUTPUT REMAINDER OF STRING 2, STRING 1 IS EXHAUSTED           DISC4550
      DO 1290 I=J1B,J2B,2                                            DISC4560
      K=K+2                                                          DISC4570
      JTAB(K)=ITAB(I)                                                DISC4580
 1290 JTAB(K+1)=ITAB(I+1)                                            DISC4590
      GO TO 1370                                                     DISC4600
C     SWITCH FROM STRING 1 OUTPUT TO STRING 2 OUTPUT                DISC4610
C     OUTPUT FROM SECOND STRING UNTIL ITEMP IS NOT EXCEEDED OR EQUALED DISC4620
 1300 J1A=I                                                          DISC4630
      ITEMP=ITAB(I)                                                  DISC4640
      DO 1320 I=J1B,J2B,2                                            DISC4650
      IF(ITAB(I)-ITEMP) 1350,1350,1310                              DISC4660
 1310 K=K+2                                                          DISC4670
      JTAB(K)=ITAB(I)                                                DISC4680
 1320 JTAB(K+1)=ITAB(I+1)                                            DISC4690
C     OUTPUT REMAINDER OF STRING 1, STRING 2 IS EXHAUSTED           DISC4700
 1330 DO 1340 I=J1A,J2A,2                                            DISC4710
      K=K+2                                                          DISC4720
      JTAB(K)=ITAB(I)                                                DISC4730
 1340 JTAB(K+1)=ITAB(I+1)                                            DISC4740
      GO TO 1370                                                     DISC4750
C     SWITCH FROM STRING 2 OUTPUT TO STRING 1 OUTPUT                DISC4760
 1350 J1B=I                                                          DISC4770
 1360 ITEMP=ITAB(I)                                                  DISC4780
      GO TO 1260                                                     DISC4790
 1370 IF(J2-J2B) 1380,1380,1230                                     DISC4800
C     COPY JTAB BACK INTO ITAB                                      DISC4810
 1380 K=J1                                                           DISC4820
      DO 1390 I=1,NSB2                                               DISC4830
      ITAB(K)=JTAB(I)                                                DISC4840
 1390 K=K+1                                                          DISC4850
C     DOUBLE LENGTH OF MERGED STRING AND TEST FOR COMPLETION        DISC4860
      IDBL=IDBL+IDBL                                                 DISC4870
      IF(NSB2-IDBL) 1400,1400,1220                                  DISC4880
C     EXIT THE SORTING ROUTINE                                      DISC4890
 1400 CONTINUE                                                       DISC4900
      RETURN
      END
```

166

```
      SUBROUTINE COMPT
      COMMON /THTA/ THETA,FR(2),VPC(2),TN(2)
      DO 20 K=1,2
      VPC(K)=0.0
      IF(FR(K).EQ.0.0) GO TO 20
      VPC(K)=(FR(K)/TN(K))
   20 CONTINUE
      IF(FR(1).NE.0.0) GO TO 30
      THETA = -VPC(2)*100.0
      GO TO 10
   30 IF(FR(2).NE.0.0) GO TO 40
      THETA = 100.0 + (VPC(1)*100.0)
      GO TO 10
   40 THETA=VPC(1)/VPC(2)
   10 RETURN
      END
      SUBROUTINE INDEX
      COMMON NVAR,JND(50),KND,IVAL(50),MAXVAL(50),IND,INDICS(50,500)
      INTEGER*2 MAXVAL,IND,INDICS
      INTEGER*2 IVAL,JND,KND
      IF(NVAR.EQ.1) GO TO 2
      MND=IVAL(1)
      DO 1 I=2,NVAR
      MAX=MAXVAL(I)
      IVL=IVAL(I)
    1 MND=MAX*(MND-1)+IVL
      IND=MND
      RETURN
    2 IND=IVAL(1)
      RETURN
      END
      SUBROUTINE DINDEX
      COMMON NVAR,JND(50),KND,IVAL(50),MAXVAL(50),IND,INDICS(50,500)
      REAL*4 MAX,K
      INTEGER*2 MAXVAL,IND,INDICS
      INTEGER*2 IVAL,JND,KND
      IF(NVAR.EQ.1) GO TO 4
      L=NVAR+1
      XND=IND
      DO 1 I=2,NVAR
      L=L-1
      MAX=MAXVAL(L)
      K=XND/MAX
      CALL REALVL   (K)
      XVL     =XND-K*MAX
      IVAL(L)=XVL
      IF(IVAL(L)) 2,2,3
    2 IVAL(L)=MAXVAL(L)
      GO TO 1
    3 K=K+1.
    1 XND=K
    4 IVAL(1)=XND
      RETURN
      END
      SUBROUTINE CHECK (IP)
      COMMON NVAR,JND(50),KND,IVAL(50),MAXVAL(50),IND,INDICS(50,500)
      INTEGER*2 MAXVAL,IND,INDICS
      INTEGER*2 IVAL,JND,KND
      MAX=JND(IP)
      IF(JND(IP).EQ.0) GO TO 2
```

167

```
      DO 1 I=1,MAX
      KND=I
      IF(IND.EQ.INDICS(IP,I)) GO TO 3
    1 CONTINUE
    2 MAX=MAX+1
      JND(IP)=MAX
      KND=MAX
      IF(KND.LE.500) GO TO 4
      WRITE (6,500)
  500 FORMAT ('0THE PROGRAM LIMIT FOR THE MAXIMUM NUMBER OF RESPONSE PAT
     ATERNS (500) HAS BEEN EXCEEDED. JOB ABORTED.')
      STOP
    4 INDICS(IP,MAX)=IND
    3 RETURN
      END
      SUBROUTINE REALVL   (/K/)
      REAL*4 K
      YK=K/1.0E5
      JK=YK
      YK=JK
      YK=YK*1.0E5
      XK=K-YK
      JK=XK
      XK=JK
      K=YK+XK
      RETURN
      END
```

6.7 PROG. VARSEL

Description: This is a variable-selection program that utilizes a distributional property of the Kullback [1959] minimum information divergence statistic to provide stopping rules for the inclusion of variables and to determine the contribution of a new variable in the presence of the levels of those variables already in the system. The program begins by ordering the random variables X_1, X_2, ..., X_p on the basis of the magnitude of all individually calculated divergences. The variable with the largest divergence is selected first. The procedure continues by observing the best pair that includes the first variable selected. If this pair is significant, then the two conditional divergences $J(X_i|X_j=0)$ and $J(X_i|X_j=1)$, $i \neq j$, are computed. The program continues in a similar fashion until the inclusion of a new variable yields a result consistent with the hypothesis that the multinomial distributions so induced in the two groups are identical. Input to the program is the user's original data matrix (coded in 1, 2 form—corresponding to the levels of the variables) as well as some control parameters and optional information.

Card Preparation: The following cards are necessary for running VARSEL:

1. *Problem Card*

Columns	Contents
1–6	Alphanumeric problem name
7–8	Number of variables
9–10	Maximum number of variables for inclusion
11	Blank
12	1 if label cards are desired 0 if label cards are not desired
13–14	Critical value for determining group membership

2. *Variable Name and Level Card*

Columns	Contents
1–2	Alphanumeric—first-variable (code)
3–5	Number of levels of first variable
6–7	Alphanumeric—second-variable (code)
8–10	Number of levels of second variable

⋮

3. *Variable Label Cards*
   a. *Label Card* #1

Columns	Contents
1–20	Alphanumeric label for first variable
21–40	Alphanumeric label for second variable
41–60	Alphanumeric label for third variable

   b. *Label Card* #2

Columns	Contents
1–20	Alphanumeric label for fourth variable
⋮	⋮
41–60	Alphanumeric label for sixth variable

c. *Label Card* #3

Columns	Contents
1–20	Alphanumeric label for seventh variable

$\vdots$ $\qquad$ $\vdots$

4. χ^2 *Critical Values Card* ($\alpha = *$)

Columns	Contents
1–6	χ^2 Critical value with 1 dif.
1–6	χ^2 Critical value with 1 dif.
7–12	χ^2 Critical value with 3 dif.
13–18	χ^2 Critical value with 7 dif.

The degree of freedom with k variables is given by (LEVEL (1)*LEVEL (2)*LEVEL (3)*...*LEVEL (k) − 1). The level of significance, α, is set by the user.

5. *Variable Format Card.* Using 60 columns specify the location of each variable using I format. For example, the variable format card used in the program listing is (I2, 9II). Note that the first two fields correspond to IG, which is the value of the group membership variable.

6. *Data Cards.* The data deck must be punched according to the variable format specified by the **VARIABLE FORMAT CARD**. Note that the predictor variables must be coded in 1, 2 form.

7. *Finish Card*

Columns	Contents
1–6	FINISH

The arrangement of cards within the deck is as shown in the diagram on the opposite page.

Output: The output of VARSEL is as follows:

1. The number of variables and their levels, variable labels, along with the sample sizes and induced distributions for each group, are presented first.
2. Next, at each step, the program prints out the group and conditional divergences as well as the appropriate χ^2 critical values.

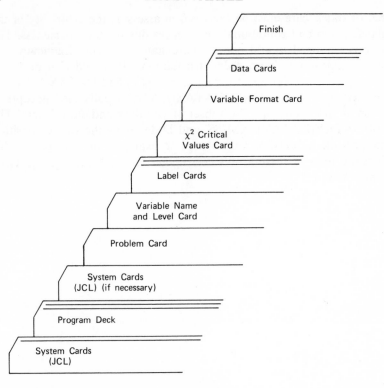

3. The number and percent misclassifications at each step are also reported.
4. Finally, all of the lower-order frequency distributions for the selected variables are presented.

Uses and Limitations: This program is designed to accommodate up to nine variables each having two levels. It is a simple matter, however, to increase the number of variables for consideration by changing those dimension statements marked by an asterisk (*). In addition, the program can be amended to handle cases where the number of levels for the variables is in excess of two. Note, however, that in large problems core requirements are likely to become severe, especially in situations where the variables have levels more than a dichotomy since the number of possible states increases dramatically, hence increasing the amount of lower-order distributions to be stored. In cases such as this, VARSEL could be modified to run in two steps. The first step (program) would determine those variables worthy of inclusion, and the second step (program) would compute the corresponding conditional divergences. Furthermore, with a proliferation of the number of states, care must be exercised in assessing the credibility of the calculated group and conditional divergences due to state sparseness. The program is equipped to add one to those states with zero frequency and adjust the appropriate lower-order distributions. It should also be noted that the method used for constructing arrays X(Y) and DATAX(DATAY) is discussed by Hartley [1962]; DATAX(DATAY) pulls the appropriate sums out of X(Y) for a specific subset of variables and stores them. The rows of DATAX(DATAY), XMISS, and D designate the storage position for all possible subsets of variables. For example, in a three-variable problem the rows of these arrays correspond to X_1, X_2, X_1X_2, X_3, X_1X_3, X_2X_3, and $X_1X_2X_3$.

```
PROG. VARSEL                          GOLDSTEIN-DILLON
A STEPWISE DISCRETE VARIABLE SELECTION PROGRAM BASED
C ON THE KULLBACK MINIMUM INFORMATION DIVERGENCE STATISTIC AS
C DICUSSED BY GOLDSTEIN AND DILLON
      DIMENSION FMT(9),II(9),HEAD(9),ISTEP(9),KOUNT(9),LASTS(9),          *
     1CHISQ(9),D(512),YYNS(512),XXNS(512),COND(256),XMISS(512)            *
      INTEGER*2 CONFIG(512,9),DATAX(512,129),DATAY(512,129),X(20000),     *
     1Y(20000)                                                           *
      DIMENSION KOUNT2(9),MSTEP(9),LOC(9),LEVEL(9),FMT1(15),LABEL(45)     *
      COMMON CONFIG
      DATA FINI/'FINI'/,SH/'SH'/
    1 FORMAT(A4,A2,I2,I2,1X,I1,I2)
    2 FORMAT(26H1SELECTION OF VARIABLES...A4,A2//)                        *
    3 FORMAT(32H0LEVELS OF FACTORS........LABELS/)                        *
    4 FORMAT(3X,A2,7X,I4,10X,5A4)
    6 FORMAT(2X,8HRESPONSE10X,6HGROUP15X,6HGROUP2/2X,7HPATTERN11X,         *
     16HCOUNTS5X,6HCOUNTS/)
    7 FORMAT(9(A2,I3))
    8 FORMAT(1H ,9I1,4X,I3,3X,I6,5X,I6)
    9 FORMAT(9H1VARIABLE12X,8HRESPONSE1X,10HDIVERGENCE1X,8HCRITICAL3X,     *
     16HNUMBER4X,7HPERCENT3X,10HCONDITIONL1X,8HCRITICAL/1X,               *
     28HSELECTED13X,7HPATTERN3X,6HVALUES4X,6HVALUES2X,10HMISCLASSED1X,     *
     310HMISCLASSED1X,10HDIVERGENCE2X,6HVALUES/)
   10 FORMAT(9F6.2)
   11 FORMAT(///,21X,9I1)
   12 FORMAT('+',72X,F10.3,F8.2/21X,'-----------------------------------',   *
     1'----------------------------------------')
   13 FORMAT('1',8X,5HLOWER2X,5HORDER1X,9HFREQUENCY1X,                     *
     113HDISTRIBUTIONS//)
   14 FORMAT(1X,9HVARIABLES9X,8HRESPONSE5X,6HGROUP16X,                     *
     16HGROUP2/2X,8HSELECTED10X,7HPATTERN4X,9HFREQUENCY3X,9HFREQUENCY/)    *
   15 FORMAT(3(5A4))                                                      *
   16 FORMAT(15A4)                                                        *
   19 FORMAT('1',41X,'A STEPWISE DISCRETE VARIABLE SELECTION PROGRAM'/,
     137X,'BASED ON THE KULLBACK MINIMUM INFORMATION DIVERGENCE',
     2' STATISTIC AS'/42X,'DISCUSSED BY GOLDSTEIN AND DILLON WRITTEN',
     3' AT'/37X,'BERNARD M BARUCH COLLEGE CUNY VERSION OF JUNE 1976')
 1101 FORMAT(13H0SAMPLE SIZES2X,7HGROUP1=,F4.0/15X,7HGROUP2=,F4.0/15X,
     17HTOTAL =,F4.0//)
C READ AND PRINT CONTROL AND OPTIONAL DATA
      WRITE(6,19)
  100 READ(5,1)PR,PR1,K,MAXV,IY,IXX
      IF(PR.EQ.FINI) GO TO 7777
      GO TO 8888
 7777 IF(PR1.EQ.SH) GO TO 9999
 8888 CONTINUE
      WRITE(6,2)PR,PR1
      READ(5,7) (HEAD(I),LEVEL(I),I=1,K)
      IF(IY.EQ.0) GO TO 37
      READ(5,15)LABEL
   37 READ(5,10)CHISQ
      WRITE(6,3)
      NM=1
      LM=5
      DO 36 I=1,K
      WRITE(6,4)HEAD(I),LEVEL(I),(LABEL(J),J=NM,LM)
      NM=NM+5
      LM=LM+5
   36 CONTINUE
      N=LEVEL(1)
```

173

```
          DO 102 I=2,K
    102 N=N*LEVEL(I)
          READ(5,16)FMT1
          DO 101 I=1,N
          X(I)=0.
    101 Y(I)=0.
          DAT=0.0
C READ INPUT DATA AND COMPUTE CELL FREQUENCIES
    107 READ(9,FMT1,END=108)IG,(II(I),I=1,K)
          M=II(1)
          MM=1
          DO 103 I=2,K
          MM=MM*LEVEL(I-1)
    103 M=M+MM*(II(I)-1)
          IF(IG.GE.IXX) GO TO 106
    104 X(M)=X(M)+AMAX1(1.0,DAT)
          GO TO 107
    106 Y(M)=Y(M)+AMAX1(1.0,DAT)
          GO TO 107
    108 CONTINUE
          NXX=(2**MAXV)+1
          DO 115 I=1,N
          DATAX(I,NXX)=0
          DATAY(I,NXX)=0
    115 D(I)=0.0
          XNS=0.
          YNS=0.
C COMPUTE SAMP;E SIZES
          DO 111 I=1,N
          XNS=XNS+X(I)
    111 YNS=YNS+Y(I)
          TNS=XNS+YNS
          WRITE(6,1101)XNS,YNS,TNS
          WRITE(6,6)
          CALL PATTN(K)
C PRINT RESPONSE PATTERNS AND CELL FREQUENCIES
          DO 110 I=1,N
    110 WRITE(6,8)(CONFIG(I,J),J=1,K),I,X(I),Y(I)
          CALL AVDAT (K,LEVEL,N,X,L,ISTEP,KOUNT,Y)
          CALL AVCAL (K,LEVEL,X,L,ISTEP,LASTS,Y)
          CALL MEANQ(K,LEVEL,X,GMEAN,ISTEP,KOUNT,LASTS,Y,DATAX,DATAY,
         1           D,XNS,YNS,XMISS,MAXV,NXX)
C DETERMINE FIRST VARIABLE TO ENTER
          ISTEP(1)=1
          KOUNT(1)=0
          DO 34 I=2,K
          KOUNT(I)=0
     34 ISTEP(I)=ISTEP(I-1)*LEVEL(I)
          KK=1
          XHOLD=-9999
          DO 35 I=1,K
          IF(D(ISTEP(I)).LT.XHOLD) GO TO 35
          XHOLD=D(ISTEP(I))
          LOC(KK)=ISTEP(I)
          KOUNT(KK)=I
     35 CONTINUE
          WRITE(6,9)
          CALL PRINT(K,KK,KOUNT,CHISQ,XHOLD,XMISS,HEAD,TNS,LOC)
C DETERMINE REMAINING VARIABLES TO ENTER AND CONDITIONAL DIVERGENCES
   1000 KK=KK+1
```

```
      IF(KK.GT.MAXV) GO TO 99
      CALL SELECT(K,KK,CHISQ,XHOLD,KOUNT,ISTEP,LOC,D)
      CALL PRINT(K,KK,KOUNT,CHISQ,XHOLD,XMISS,HEAD,TNS,LOC)
      IF(XHOLD.LE.CHISQ(KK)) GO TO 99
      CALL CONDL(KK,KOUNT,ISTEP,DATAX,DATAY,LL,IND,LOC,COND,NXX)
      CALL PATTN(KK)
C COLLECT PROPER RESPONSE PATTERNS FOR CONDITIONALS
      I=1
      IJK=1+IND
      DO 50 IB=1,LL
      IF(I.NE.IJK) GO TO 52
      I=IJ+1
      IJK=I+IND
   52 CONTINUE
      IJ=I+IND
      DO 53 J=I,IJ,IND
      WRITE(6,11)(CONFIG(J,L),L=1,KK)
   53 CONTINUE
      WRITE(6,12)COND(IB),CHISQ(1)
      I=I+1
   50 CONTINUE
      GO TO 1000
   99 CONTINUE
      WRITE(6,13)
      WRITE(6,14)
      CALL FREQ(K,KK,LOC,DATAX,DATAY,HEAD,KOUNT)
      GO TO 100
 9999 CONTINUE
      WRITE(6,999)
  999 FORMAT('1','FINISH')
      STOP
      END
C COMPUTES RESPONSE PATTERNS
      SUBROUTINE PATTN(N)
      INTEGER*2 CONFIG(512,9)                                       *
      COMMON CONFIG
      NN=N
      L=1
   20 J=1
      S=1
   10 CONFIG(J,NN)=0
      IF(J-(2**(N-L))*(2*S-1))1,2,1
    1 J=J+1
      GO TO 10
    2 J=J+1
    4 CONFIG(J,NN)=1
      IF(J-(2**(N-L))*(2*S))5,6,5
    5 J=J+1
      GO TO 4
    6 IF(S-2**(L-1))7,8,7
    7 S=S+1
      GO TO 1
    8 IF(L-N)11,12,11
   11 L=L+1
      NN=NN-1
      GO TO 20
   12 CONTINUE
      RETURN
      END
C PRINTS SUMMARY STATISTICS
```

```
      SUBROUTINE PRINT(K,KK,KOUNT,CHISQ,XHOLD,XMISS,HEAD,TNS,LOC)
      DIMENSION KOUNT(9),CHISQ(9),XMISS(512),HEAD(9),LOC(9),          *
     1           MSTEP(9),FMT(9)                                      *
      L=0
      DO 50 I=1,K
   50 MSTEP(I)=0
   60 L=L+1
      IF(L.GT.KK) GO TO 67
      DO 65 I=1,K
      IF(MSTEP(I).EQ.1) GO TO 65
      IF(I.NE.KOUNT(L)) GO TO 65
      MSTEP(I)=1
      GO TO 60
   65 CONTINUE
   67 L=0
      DO 75 I=1,K
      FMT(I)=BLANK
      IF(MSTEP(I))75,75,80
   80 L=L+1
      FMT(L)=HEAD(I)
   75 CONTINUE
      PCT=XMISS(LOC(KK))*100/TNS
      WRITE(6,2)(FMT(I),I=1,9),XHOLD,CHISQ(KK),XMISS(LOC(KK)),PCT
    2 FORMAT(1X,9A2,11X,F10.5,1X,F8.2,F11.0,F11.2/)
      RETURN
      END
C SELECTION OF VARIABLES TO ENTER
      SUBROUTINE SELECT(K,KK,CHISQ,XHOLD,KOUNT,ISTEP,LOC,D)
      DIMENSION CHISQ(9),KOUNT(9),ISTEP(9),LOC(9),D(512)             *
      XHOLD=CHISQ(KK)
      JX=KK-1
    1 DO 50 I=1,K
      DO 51 IK=1,JX
      IF(I.EQ.KOUNT(IK)) GO TO 50
   51 CONTINUE
      INDEX=LOC(KK-1)+ISTEP(I)
      IF(D(INDEX).LT.XHOLD) GO TO 50
      XHOLD=D(INDEX)
      LOC(KK)=INDEX
      KOUNT(KK)=I
   50 CONTINUE
      IF(XHOLD.NE.CHISQ(KK)) GO TO 100
      XHOLD=-9999
      GO TO 1
  100 RETURN
      END
C COMPUTES CONDITIONAL DIVERGENCES
      SUBROUTINE CONDL(KK,KOUNT,ISTEP,DATAX,DATAY,LL,IND,LOC,COND,NXX)
      DIMENSION KOUNT(9),ISTEP(9),LOC(9),COND(256),KOUNT2(9)          *
      INTEGER*2 DATAX(512,129),DATAY(512,129)                         *
      JX=KK-1
      DO 25 I=1,KK
   25 KOUNT2(I)=KOUNT(I)
      DO 26 IK=1,JX
      DO 26 JK=IK,JX
      IF(KOUNT2(IK).LE.KOUNT2(JK)) GO TO 26
      IHOLD=KOUNT2(IK)
      KOUNT2(IK)=KOUNT2(JK)
      KOUNT2(JK)=IHOLD
   26 CONTINUE
```

```
      DO 27 IK=1,JX
      IF(KOUNT(KK).LT.KOUNT2(IK)) GO TO 30
   27 CONTINUE
      IK=KK
   30 IND=ISTEP(IK)
      TSUMX=0.0
      TSUMY=0.0
      I=1
      LL=DATAX(LOC(KK-1),NXX)
      IJK=1+IND
      DO 100 IB=1,LL
      UNX=DATAX(LOC(KK-1),IB)
      UNY=DATAY(LOC(KK-1),IB)
      IF(I.NE.IJK) GO TO 133
      I=IJ+1
      IJK=I+IND
  133 CONTINUE
      IJ=I+IND
      SUMX=0.0
      SUMY=0.0
      DO 131 J=I,IJ,IND
      FNX=DATAX(LOC(KK),J)
      FNY=DATAY(LOC(KK),J)
C CHECKS FOR ZERO CELL FREQUENCIES
      IF(FNX.LE.0.0)FNX=1.0
      IF(FNY.LE.0.0)FNY=1.0
      SUMX=SUMX+FNX
      SUMY=SUMY+FNY
      TSUMX=TSUMX+FNX
      TSUMY=TSUMY+FNY
  131 CONTINUE
C ADJUSTS CELLS FOR ZERO FREQUENCIES
      UNX=SUMX
      UNY=SUMY
      COND(IB)=0.0
      DO 132 J=I,IJ,IND
      FNX=DATAX(LOC(KK),J)
      FNY=DATAY(LOC(KK),J)
      IF(FNX.LE.0.0)FNX=1.0
      IF(FNY.LE.0.0)FNY=1.0
      ZZ=(UNY/UNX)
      COND(IB)=COND(IB)+(((FNX/UNX)-(FNY/UNY))*ALOG((FNX/FNY)*ZZ))
  132 CONTINUE
      I=I+1
  100 CONTINUE
      DO 101 IB=1,LL
  101 COND(IB)=((TSUMX*TSUMY)/(TSUMX+TSUMY))*COND(IB)
      RETURN
      END
C COMPUTES DIVERGENCES FOR ALL SUBSETS
      SUBROUTINE MEANQ(K,LEVEL,X,GMEAN,MSTEP,KOUNT,LASTS,Y,DATAX,
     1                 DATAY,D,XNS,YNS,XMISS,MAXV,NXX)
      INTEGER*2 X(20000),Y(20000),DATAX(512,129),DATAY(512,129)    *
      DIMENSION LEVEL(9),MSTEP(9),KOUNT(9),LASTS(9),D(512),XXNS(512), *
     1YYNS(512),XMISS(512)                                         *
      NX=(2**MAXV)
      N=LEVEL(1)
      DO 150 I=2,K
  150 N=N*LEVEL(I)
      LASTS(1)=LEVEL(1)
```

177

```
      DO 178 I=2,K
  178 LASTS(I)=LEVEL(I)+1
      NN=1
      LL=(2**K)-1
      MSTEP(1)=1
      DO 180 I=2,K
  180 MSTEP(I)=MSTEP(I-1)*2
      DO 185 I=1,LL
      XXNS(I)=XNS
      YYNS(I)=YNS
  185 XMISS(I)=0.0
      DO 190 I=1,K
  190 KOUNT(I)=0
  200 L=0
      DO 260 I=1,K
      IF(KOUNT(I)-LASTS(I)) 210, 250, 210
  210 IF(L) 220, 220, 240
  220 KOUNT(I)=KOUNT(I)+1
      IF(KOUNT(I)-LEVEL(I)) 230, 230, 250
  230 L=L+MSTEP(I)
      GO TO 260
  240 IF(KOUNT(I)-LEVEL(I)) 230, 260, 230
  250 KOUNT(I)=0
  260 CONTINUE
      IF(L) 285,285,270
  270 IVAL=DATAX(L,NXX)
      IF(IVAL.GE.NX) GO TO 25
      IVAL=IVAL+1
      DATAX(L,IVAL)=X(NN)
      DATAY(L,IVAL)=Y(NN)
      DATAX(L,NXX)=IVAL
      DATAY(L,NXX)=IVAL
   25 CONTINUE
      A=X(NN)
      B=Y(NN)
C COMPUTES THE NUMBER OF MISCLASSIFICATIONS
      XMISS(L)=XMISS(L)+AMIN1(A,B)
C CHECKS AND ADJUSTS FOR ZERO CELLS
      IF(A) 26,26,27
   26 A=A+1
      XXNS(L)=XXNS(L)+1
   27 IF(B) 28,28,30
   28 B=B+1
      YYNS(L)=YYNS(L)+1
   30 PFX=A/XXNS(L)
      PFY=B/YYNS(L)
C COMPUTES DIVERGENCES
      D(L)=D(L)+((PFX-PFY)*ALOG((A/B)*(YYNS(L)/XXNS(L))))
      NN=NN+1
      GO TO 200
  285 GMEAN=X(NN)+Y(NN)
      DO 310 I=1,LL
  310 D(I)=D(I)*((XXNS(I)*YYNS(I))/(XXNS(I)+YYNS(I)))
      RETURN
      END
```

References

Anderson, J. A. [1972]. "Separate sample logistic discrimination," *Biometrika*, **59**, 19–36.

Anderson, T. W. [1951]. "Classification by multivariate analysis," *Psychometrika*, **16**, 31–50.

Anderson, T. W. [1973]. "Asymptotic evaluation of the probabilities of misclassification," in *Discriminant Analysis and Applications*, T. Cacoullos (Ed.), New York: Academic, pp. 17–36.

Anderson, T. W. [1952]. *An Introduction to Multivariate Statistical Methods*, New York: Wiley.

Bahadur, R. R. [1961]. "A representation of the joint distribution of response to n dichotomous items," in *Studies in Item Analysis and Prediction*, H. Solomon (Ed.), Palo Alto, Calif.: Stanford Univ. Press, pp. 158–168.

Bendel, R. B. and A. A. Afifi [1977]. "Comparison of stopping rules in forward stepwise regression," *J. Am. Stat. Assoc.*, **72** (357), 46–53.

Berkson, J. T. [1955]. "Maximum likelihood and minimum χ^2 estimation of the logistic function," *J. Am. Stat. Assoc.*, **50**, 130–162.

Bishop, Y. M. M., S. E. Fienberg, and P. W. Holland [1975]. *Discrete Multivariate Analysis: Theory and Practice*, Cambridge, Mass.: MIT Press.

Cacoullos, T. [1966]. "Estimation of a multivariate density," *Ann. Inst. Stat. Math. (Tokyo)*, **18**, 179–189.

Chang, P. C. and Afifi, A. A. [1972]. "Classification based on dichotomous and continuous variables," *J. Am. Stat. Assoc.*, **67**, 336–339.

Cochran, W. G. [1962]. "On the performance of the linear discriminant function (report on a discussion of a paper by W. G. Cochran)," *Bull. Internat. Stat. Inst.*, **35**, 157–158. [Reprinted in *Technometrics*, **6** (1964), 179–190.]

Cochran, W. G. and Bliss, C. I. [1948]. "Discriminant functions with covariance," *Ann. Math. Stat.*, **19**, 151.

Cochran, W. G. and C. E. Hopkins, [1961]. "Some classification problems with multivariate qualitative data," *Biometrics*, **17**, 10–32.

Cox, D. R. [1970]. *The Analysis of Binary Data*, London: Methuen.

Dash, J. F., L. Schiffman, and C. Berenson, [1976]. "Information search and store choice," *J. Adv. Res.*, **16**, 35–40.

Day, N. E. and D. F. Kerridge, [1967]. "A general maximum likelihood discriminant," *Biometrics*, **23**, 313–323.

Dillon, W. R. and M. Goldstein, [1978]. "On the performance of some multinomial classification rules," *J. Am. Stat. Assoc.*, **78**, no. 362.

Dillon, W. R., M. Goldstein, and L. Schiffman, [1978]. "On the appropriateness of linear discriminant and multinomial classification analysis in marketing research," *J. Mark. Res.*, **15**, 103–112.

Dunn, O. J. [1971]. "Some expected values for probabilities of correct classification in discriminant analysis," *Technometrics*, **13**, 345.

Elashoff, J. D., R. M. Elashoff, and G. E. Goldman, [1967]. "On the choice of variables in classification problems with dichotomous variables," *Biometrika*, **54**, 668.

Fienberg, S. [1977]. *The Analysis of Cross-Classified Categorical Data*, Cambridge, Mass.: MIT Press.

Fisher, R. A. [1936]. "The use of multiple measurements in taxonomic problems," *Ann. Eugenics*, **7**, 179–188.

Gilbert, E. S. [1968]. "On discrimination using qualitative variables," *J. Am. Stat. Assoc.*, **63**, 1399.

Gilbert, E. S. [1969]. "The effect of unequal variance–covariance matrices on Fisher's linear discriminant function," *Biometrics*, **25**, 505–516.

Glick, N. [1972]. "Sample-based classification procedures derived from density estimators," *J. Am. Stat. Assoc.*, **67**, 166–122.

Glick, N. [1973]. "Sample-based multinomial classification," *Biometrics*, **29**, 241–256.

Goldstein, M. [1975]. "Comparison of some density estimate classification procedures," *J. Am. Stat. Assoc.*, **70**, 666–669.

Goldstein, M. [1976]. "An approximate test for comparative discriminatory power," *Multivariate Behav. Res.*, **11**, 157–163.

Goldstein, M. and W. R. Dillon, [1977]. "A stepwise discrete variable selection procedure," *Commun. Stat., Theory and Materials, 6, 1423–36.*

Goldstein, M. and M. Rabinowitz, [1975]. "Selection of variates for the two-group multinomial classification problem," *J. Am. Stat. Assoc.*, **70**, 776–781.

Goldstein, M. and E. Wolf, [1977]. "On the problem of bias in multinomial classification," *Biometrics*, **33**, 325–331.

Goldstein, M. [1977]. "A two-group classification procedure for multivariate dichotomous responses," *Multivariate Behav. Res.*, **12**, 335–346.

Haberman, S. J. [1974]. "Loglinear models for frequency tables with ordered classifications," *Biometrics*, **30**, 589–600.

Halperin, M., W. C. Blackwelder, and J. I. Verter, [1971]. "Estimation of the multivariate logistic risk function: a comparison of the discriminant and maximum likelihood approaches," *J. Chron. Dis.*, **24**, 125–158.

Hartley, H. O. [1962]. *Mathematical Methods for Digital Computers*, New York: Wiley.

Hills, M. [1966]. "Allocation rules and their error rates," *J. Roy Stat. Soc.*, **B28**, 1.

Hills, M. [1967]. "Discrimination and allocation with discrete data," *J. Roy. Stat. Soc.*, **C16**, 237–250.

Hoel, P. G. and R. P. Peterson, [1949]. "A solution to the problem of optimum classification," *Ann. Math. Stat.*, **20**, 433–438.

Kronmal, R. A. and M. Tarter, [1968]. "The estimation of probability densities and cumulatives by Fourier series methods," *J. Am. Stat. Assoc.*, **63**, 925–952.

Krzanowski, W. J. [1975]. "Discrimination and classification using both binary and continuous variables," *J. Am. Stat. Assoc.*, **70**, 782–790.

Krzanowski, W. J. [1976]. "Canonical representation of the location model for discrimination or classification," *J. Am. Stat. Assoc.*, **71**, 845–848.

Krzanowski, W. J. [1977]. "The performance of Fisher's linear discriminant function under non-optimal conditions," *Technometrics*, **19**, 191–200.

Kullback, S. [1959]. *Information Theory and Statistics*, New York: Wiley.

Lachenbruch, P. A. [1965]. "Estimation of error rates in discriminant analysis," Ph.D. dissertation, University of California at Los Angeles.

Lachenbruch, P. A. [1966]. "Discriminant analysis when the initial samples are misclassified," *Technometrics*, **8**, 657.

Lachenbruch, P. A. [1975]. *Discriminant Analysis*, Hafner Press, New York.

Lachenbruch, P. A. and M. R. Mickey, [1968]. "Estimation of error rates in discriminant analysis," *Technometrics*, **10**, 1.

Lachin, J. M. [1973]. "On a stepwise procedure for two populations Bayes decision rules using discrete variables," *Biometrics*, **29**, 551–564.

Martin, D. C. and R. A. Bradley, [1972]. "Probability models estimation and classification for multivariate dichotomous populations," *Biometrics*, **28**, 203–222.

Matusita, K. (1954). "On estimation by the minimum distance method," *Ann. Inst. Stat. Math.*, **7**, 67–77.

Matusita, K. (1955). "Decision rules based on the distance for problems of fit, two samples and estimation," *Ann. Math. Stat.* **26**, 631–640.

Matusita, K. (1957). "Classification based on distance in multivariate Gaussian cases," *Proc. Fifth Berkeley Symp. Math. Stat. and Prob.*, **1**, 299–304.

McLachlan, G. J. [1976]. "The bias of the apparent error rate in discriminant analysis," *Biometrika*, **63**, 239–244.

Miller, R. G. [1974]. "The jackknife—a review," *Biometrika*, **61**, 1–15.

Moore, D. H., II [1973]. "Evaluation of five discrimination procedures for binary variables," *J. Am. Stat. Assoc.*, **68**, 339–404.

Morrison, D. G. [1969]. "On the interpretation of discriminant analysis," *J. Market. Res.*, **6**, 156–163.

Okamoto, M. [1963]. "An asymptotic expansion for the distribution of the linear discriminant function," *Ann. Math. Stat.*, **34**, 1286–1301.

Ott, J. and R. A. Kronmal, [1976]. "Some classification procedures for multivariate binary data using orthogonal functions," *J. Am. Stat. Assoc.*, **71**, 391–399.

Parzen, E. [1962]. "On estimation of a probability density function and mode," *Ann. Math. Stat.*, **33**, 1065–1076.

Raiffa, H. [1961]. "Statistical decision theory approach to item selection for dichotomous test and criterion variables," in *Studies in Item Analysis and Prediction*, H. Solomon (Ed.), Palo Alto, Calif.: Stanford University Press, 187–200.

Reis, T. N. and H. T. Smith, [1963]. "The use of Chi-square for preference testing in multidimensional problems," *Chem. Eng. Progr.*, **59**, 39–43.

Rosenblatt, M. [1956]. "Remarks on some non-parametric estimates of a density function," *Ann. Math. Stat.*, **27**, 832–837.

Shum, Y. Y. and A. R. Elliot, [1972]. "Computation of the fast Hadamard transform," in *Proceedings of the Third Walsh Functions Symposium*, R. W. Zeek and A. E. Showalter (Eds.), Dept. of Commerce, Springfield, Va. 22151, 177–180.

Simon, G. [1974]. "Alternative analysis for singly-ordered contingency tables," *J. Am. Stat. Assoc.*, **69**, 971–976.

Smith, C. A. B. [1947]. "Some examples of discrimination," *Ann. Eugen.*, **18**, 272–283.

Solomon, H. (Ed.) [1961]. *Studies in Item Analysis and Prediction*, Palo Alto, Calif.: Stanford University Press.

Sorum, M. [1971]. "Estimating the conditional probability of misclassification," *Technometrics*, **13**, 333.

Truett, J., J. Cornfield, and W. Kannel, [1967]. "A multivariate analysis of the risk of coronary heart disease in Framingham," *J. Chron. Dis.*, **20**, 511–524.

Van Ryzin, J. [1966]. "Bayes risk consistency of classification procedures using density estimation," *Sankhya*, **A28**, 261–270.

Weiner, J. and O. J. Dunn, [1966]. "Elimination of variates in linear discrimination problems," *Biometrics*, **22**, 268.

Welch, B. L. [1939]. "Note on discriminant functions," *Biometrika*, **31**, 218–220.

Yerushalmy, et al. [1965]. "Birth weight and gestation as indices of 'immaturity,'" *Am. J. Dis. Childr.*, **109**, 43–57.

Index